Hooray for Multiplication Facts!

by
Becky Daniel

illustrated by Judy Hierstein

Math Facts Wheel by Kenneth Holland

Cover by Judy Hierstein

Copyright © Good Apple, Inc., 1990

ISBN No. 0-86653-519-5

Printing No. 9876

GOOD APPLE, INC.
BOX 299
CARTHAGE, IL 62321

Table of Contents

GA1136

To the Teacher

Mastering the multiplication facts doesn't have to be a boring, tedious process, not if you use *Hooray for Multiplication Facts!* The puzzles, mazes, codes, magic tricks, games, learning aids and math facts wheel in this book were carefully designed to teach the basic multiplication facts while entertaining and delighting your mathematicians. Every page is a different activity format so children never get bored.

It is highly recommended that you begin this book by giving each student the three-page multiplication facts test found on pages 60, 61, and 62. Allow exactly seven minutes for the students to answer as many of the multiplication problems as they can. Record the initial scores. As the children are learning the facts, retest them weekly using the same three-page test and recording the scores each time. You may want the children to graph their own progress. (See graph on page 63.) The test can also be used as a multiplication facts aid. Simply cut along the solid lines to create mini flash cards so children can review the multiplication facts with which they are experiencing difficulty. Each child should also make his/her own multiplication facts wheel. See reproducible patterns on pages 70 and 71.

The many activity sheets found in *Hooray for Multiplication Facts!* were carefully designed and sequenced so that each page adds new facts and reviews those already learned. With the exception of several activities in the back of the book, no carrying is necessary. If children master the basic facts, mathematics is enjoyable; however, if they do not master those facts, mathematics will soon become difficult and frustrating. Don't let one single student leave your classroom this year not knowing the multiplication facts. It's fun and it's easy to sharpen math skills with the colossal collection of ideas found herein.

GA1136

Using the Math Facts Wheel

The self-teaching math wheel is a fantastic teaching tool for grades two, three and four. Students using this simple device learn the basic math facts without consuming valuable class teaching time. The teacher is freed from forcing exercises upon students because students actually enjoy the math wheel. Learning multiplication facts becomes fun. The embarrassment that sometimes accompanies the use of flash cards is eliminated. To use the wheel, the student first gives his answer to himself, then advances the wheel slightly and presto! the correct answer appears! If he was correct, seeing the correct answer will reinforce the student's memory or confidentially correct his error. Moving through the full range of basic multiplication facts on the wheel, the student rapidly achieves readiness to perform the basic math functions.

This math tool has demonstrated its usefulness with special students who are having learning difficulties. Slow learners require continuing reinforcement and correction. The math wheel meets these needs. Continued practice day after day helps these students achieve using the wheel without the teacher's constant supervision or assistance.

Assembly of the Math Facts Wheel

1. Punch out the two die cut wheels.
2. Punch out the small answer box sections from the smaller (top) wheel.
3. Using a metal brad fastener, fasten the wheels together with the smaller wheel on top. Spread the brad tips behind the large (bottom) wheel.

Assembly of Teacher-Produced Wheels

1. Reproduce a copy of each of the components of the wheel for each of your students.
2. Glue these copies to oaktag or poster board.
3. Cut out, using heavy scissors or sharp knife. (Do not allow students to use knives.)
4. Cut out answer windows with X-acto knife. It is important to stay within the windows when cutting.
5. Assemble as with steps used on original wheel (see instructions above).

How to Use the Math Facts Wheel

1. Hold the assembled wheel in both hands arranging the perimeter numbers of both wheels so that they match up.
2. For example, for the multiplication wheel, first move the wheel so that the 3 printed on its edge is directly below the 1 on the edge of the bottom wheel. The student should then think "3 × 1 = 3." Now move the top wheel slightly clockwise. The answer 3 should appear in the answer window to reinforce or correct the student.
3. Continue moving the top wheel clockwise until its number 3 is directly below the 2 on the bottom wheel. Give your answer. Move the top wheel slightly clockwise. The answer 6 will appear.
4. Continue in this fashion with the 3's going up to 3 × 12. Then proceed with the 4, applying it to each factor on the bottom wheel, etc.

GA1136

Multiplying by Zero

When you multiply a number by zero, the product is always zero. For example, if you have zero dimes in each of your pockets, no matter how many pockets you have, you have zero dimes. Complete the multiplication problems below.

1. $0 \times 1 =$ $0 \times 6 =$

2. $0 \times 3 =$ $0 \times 2 =$

3. $0 \times 4 =$ $0 \times 7 =$

4. $0 \times 5 =$ $0 \times 1000 =$

Draw a picture for each multiplication story problem below.

1. Elizabeth has two baskets. Each basket contains zero apples. How many apples in all does Elizabeth have?

2. John has three boxes. Each box contains zero pieces of candy. How many pieces of candy in all does John have?

Bonus: Which weighs more, zero baskets with ten apples in each basket or ten baskets with zero apples in each?

_____Multiplying by One_____

Complete each picture and write a multiplication sentence for each story problem below. The first picture has been completed for you.

---

1. Diane has two baskets of apples and each basket contains one apple. How many apples in all does Diane have?

2 × 1 = 2

---

2. Jack has three boxes of candy and each box contains one piece of candy. How many pieces of candy in all does Jack have?

---

3. Dick has three coins in each of his pockets. He has a total of one pocket. How many coins in all does Dick have?

---

4. Ruth put her collection of twenty-seven marbles all in one bag. How many marbles in all does Ruth have?

TIME AFTER TIME WITH ME YOU GET NOWHERE FAST!

Bonus: Write a multiplication story problem and draw a picture for it.

GA1136

Multiplying by Two

Draw dots on each side of the dominoes below to show the multiplication problems. The first one has been completed for you.

1. 2 × 1 =

2. 2 × 2 =

3. 2 × 3 =

4. 2 × 5 =

5. 2 × 4 =

6. 2 × 6 =

7. 2 × 0 =

Draw dots on the dominoes below to show the multiplication problems. The first one has been completed for you.

1. 3 × 2 =

2. 4 × 2 =

3. 2 × 2 =

4. 5 × 2 =

Bonus: Draw dominoes to show this problem: 6 × 2 = 12.

Multiplication Line Design

Use a ruler to draw a line connecting each number with the number that is exactly two times as big. Example: Connect 2 and 4, 3 and 6, etc. First connect the appropriate inside numbers. Then connect the appropriate outside numbers.

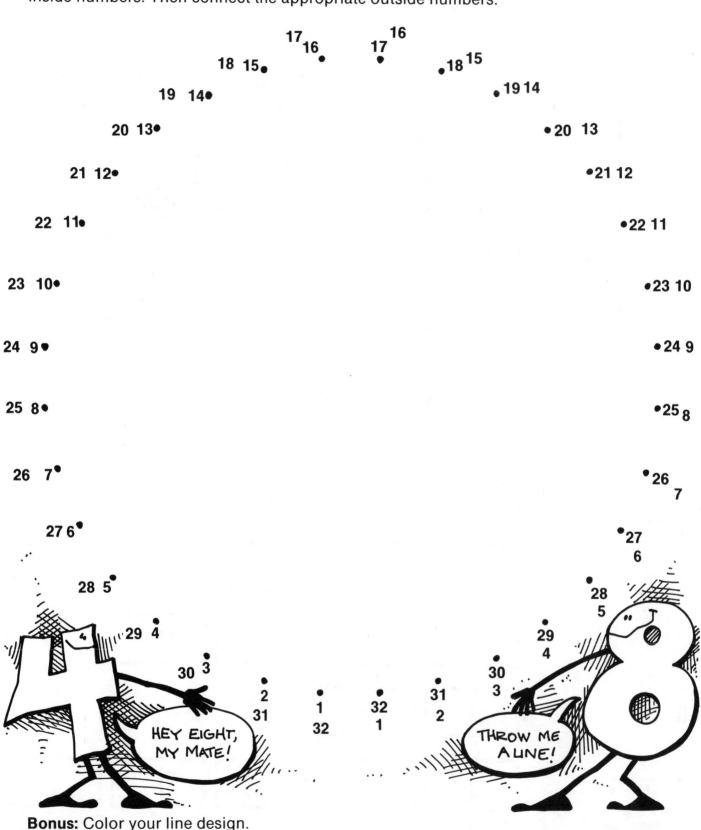

Bonus: Color your line design.

4

Multiplying by Three

Write a multiplication problem for each picture below. The first picture has been completed for you.

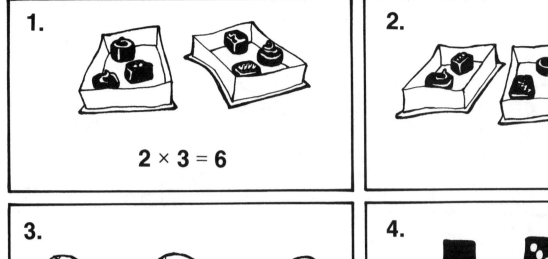

1.

$2 \times 3 = 6$

2.

3.

4.

5.

6.

7.

8.

Bonus: Which weighs more, one thousand bags each containing five pieces of gold or five bags each containing one thousand pieces of gold?

GA1136

Multiplying by Four

Draw four jelly beans in each bag to calculate the multiplication problems below. The first one has been completed for you.

1. $1 \times 4 =$

2. $2 \times 4 =$

3. $3 \times 4 =$

4. $4 \times 4 =$

5. $5 \times 4 =$

6. $6 \times 4 =$

7. $7 \times 4 =$

8. $8 \times 4 =$

9. $9 \times 4 =$

10. $10 \times 4 =$

11. $11 \times 4 =$

12. $12 \times 4 =$

ALL THESE JELLY BEANS COST A "FOUR"-TUNE.

STOP EATING THEM BE-"FOUR" YOU'RE SICK!

DON'T YOU LIKE BEANS?

THEY'RE NOT MY BAG!

Bonus: Draw bags of jelly beans to represent this problem: $4 \times 12 =$

GA1136

Multiplying by Five

Write a multiplication problem for each set of tally marks below. The first one has been completed for you.

1. 卅 卅 $2 \times 5 = 10$

2. 卅

3. 卅 卅 卅

4. 卅 卅 卅 卅

5. 卅 卅 卅 卅 卅 卅

6. 卅 卅 卅 卅 卅 卅 卅

7. 卅 卅 卅 卅 卅 卅 卅 卅

8. 卅 卅 卅 卅 卅 卅 卅 卅 卅

9. 卅 卅 卅 卅 卅 卅 卅 卅 卅 卅 卅

10. 卅 卅 卅 卅 卅

11. 卅 卅 卅 卅 卅 卅 卅 卅 卅 卅

12. 卅 卅 卅 卅 卅 卅 卅 卅 卅 卅 卅 卅

Bonus: Draw tally marks to show this multiplication problem: $5 \times 4 =$

GA1136

Multiplying by Six

A hexagon has six sides. Count sides of the hexagons to solve each multiplication problem below. The first problem has been completed for you.

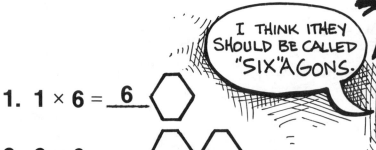

1. $1 \times 6 =$ ___6___

2. $2 \times 6 =$ ___

3. $4 \times 6 =$ ___

4. $11 \times 6 =$ ___

5. $5 \times 6 =$ ___

6. $10 \times 6 =$ ___

7. $7 \times 6 =$ ___

8. $9 \times 6 =$ ___

9. $6 \times 6 =$ ___

10. $8 \times 6 =$ ___

11. $3 \times 6 =$ ___

12. $12 \times 6 =$ ___

Bonus: What shape would you have to use to show this multiplication problem:
$6 \times 4 =$?

8

Roll 'Em

Draw the dots on each pair of dice to show the product given in the upper left-hand corner of each box. Remember that each die only has 1, 2, 3, 4, 5 or 6 dots. The first one has been completed for you.

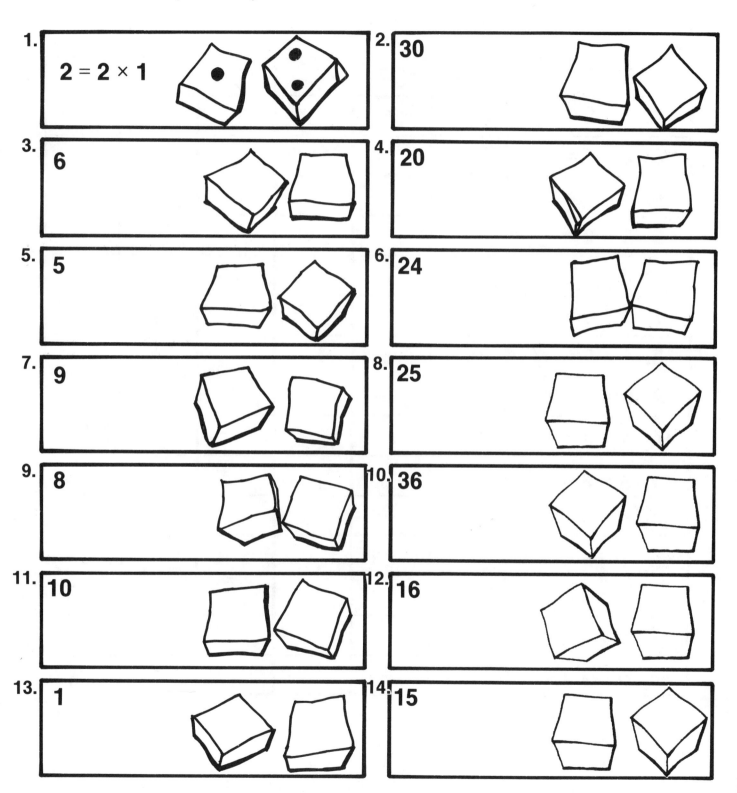

1. $2 = 2 \times 1$

2. 30

3. 6

4. 20

5. 5

6. 24

7. 9

8. 25

9. 8

10. 36

11. 10

12. 16

13. 1

14. 15

Bonus: What three products can be rolled two different ways using a pair of dice?

GA1136

Multiplication Check

Place check marks in the appropriate number of boxes in each row to calculate the product of each multiplication problem below. The first row has been completed for you.

	1	2	3	4	5	6
1. $5 \times 2 = 10$	✓✓	✓✓	✓✓	✓✓	✓✓	
2. $1 \times 2 =$						
3. $4 \times 5 =$						
4. $3 \times 5 =$						
5. $3 \times 2 =$						
6. $3 \times 4 =$						
7. $3 \times 6 =$						
8. $4 \times 6 =$						
9. $0 \times 5 =$						
10. $2 \times 6 =$						
11. $5 \times 6 =$						
12. $0 \times 3 =$						
13. $1 \times 1 =$						
14. $6 \times 4 =$						
15. $2 \times 5 =$						
16. $2 \times 4 =$						

LOOSEN UP YOUR WRISTS AND START CHECKING THIS OUT.

Bonus: Would $2 \times 6 = 12$ and $6 \times 2 = 12$ look the same on this check chart?

GA1136

Multiplying by Seven

Circle every seventh number. Then use the circled numbers to help you solve each multiplication problem below.

1	2	3	4	5	6	(7)	8	9	10	11	12
13	14	15	16	17	18	19	20	21	22	23	24
25	26	27	28	29	30	31	32	33	34	35	36
37	38	39	40	41	42	43	44	45	46	47	48
49	50	51	52	53	54	55	56	57	58	59	60
61	62	63	64	65	66	67	68	69	70	71	72
73	74	75	76	77	78	79	80	81	82	83	84

1. $7 \times 1 =$ $7 \times 4 =$ $7 \times 8 =$

2. $6 \times 7 =$ $9 \times 7 =$ $4 \times 7 =$

3. $7 \times 3 =$ $7 \times 6 =$ $7 \times 2 =$

4. $8 \times 7 =$ $1 \times 7 =$ $10 \times 7 =$

5. $7 \times 5 =$ $7 \times 10 =$ $7 \times 9 =$

6. $0 \times 7 =$ $5 \times 7 =$ $3 \times 7 =$

7. $7 \times 11 =$ $7 \times 12 =$ $7 \times 0 =$

8. $11 \times 7 =$ $12 \times 7 =$ $9 \times 7 =$

Bonus: Using the chart above, cross out every sixth number and put a box around every fifth number.

11

Multiplying by Eight

Color every eighth number yellow. Use the chart to help you solve each multiplication problem below.

1	2	3	4	5	6	7	8	9	10	11	12
13	14	15	16	17	18	19	20	21	22	23	24
25	26	27	28	29	30	31	32	33	34	35	36
37	38	39	40	41	42	43	44	45	46	47	48
49	50	51	52	53	54	55	56	57	58	59	60
61	62	63	64	65	66	67	68	69	70	71	72
73	74	75	76	77	78	79	80	81	82	83	84
85	86	87	88	89	90	91	92	93	94	95	96

1. $8 \times 1 =$ $8 \times 5 =$ $8 \times 2 =$

2. $6 \times 8 =$ $8 \times 9 =$ $8 \times 0 =$

3. $0 \times 8 =$ $3 \times 8 =$ $2 \times 8 =$

4. $5 \times 8 =$ $9 \times 8 =$ $8 \times 7 =$

5. $10 \times 8 =$ $1 \times 8 =$ $8 \times 4 =$

6. $12 \times 8 =$ $11 \times 8 =$ $7 \times 8 =$

7. $8 \times 10 =$ $4 \times 8 =$ $8 \times 12 =$

8. $8 \times 3 =$ $8 \times 6 =$ $8 \times 11 =$

EIGHT IS GREAT!

Bonus: Add the digits that you have colored in order and list each total. Example: 8 = 8, 16 = 1 + 6 = 7, 24 = 2 + 4 = 6, etc.

GA1136

Magic Squares

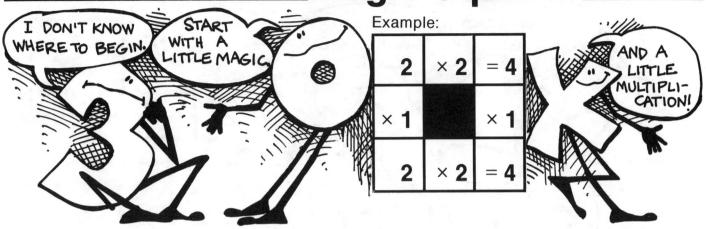

Complete each magic square. The first two numbers in each row and each column must have a product of the third number.

1.

0	6	
4	■	2

2.

2		2
	■	
2		6

3.

	2	
2	■	2
		8

4.

8		16
	■	
8		0

5.

4		8
3	■	
		0

6.

3		9
	■	1
3		

7.

5		5
	■	4
	2	

8.

	4	
6	■	3
		24

9.

	4	
2	■	1
6		

Bonus: Make up your own multiplication magic square.

GA1136

The Shape of Things

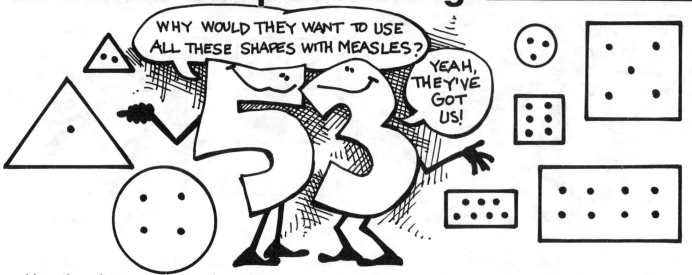

Use the shape code to decode each multiplication problem below. The first one has been completed for you.

1. ▭ × ☐ =　○ × ☐ =　○ × ☐ =

$$7 \times 5 = 35$$

2. △ × △ =　○ × ▭ =　▭ × ▭ =

3. △ × ☐ =　☐ × ◻ =　△ × ▭ =

4. △ × △ =　△ × ○ =　△ × ▭ =

5. △ × △ =　☐ × △ =　☐ × △ =

6. ○ × ▭ =　☐ × △ =　☐ × ▭ =

Bonus: Draw three multiplication problems, including product, in code.

14

Multiplying by Nine

Did you know you can use your hands as a calculator when multiplying any one-digit number by nine?

1. Hold up all ten fingers. Give each finger an imaginary number starting with the little finger on the left hand and ending with the little finger on the right hand.

2. Tuck down the finger that stands for the number you wish to multiply by nine. Four times nine would look like this:

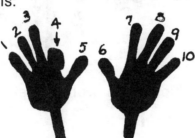

3. Count the fingers on the left side of the finger you tucked down. Those fingers stand for tens. (3 × 10 = 30) Then count the fingers on the right side of the finger you tucked down. Those fingers stand for ones. (6 × 1 = 6) Add those two to get the answer. (30 + 6 = 36)

Use your finger calculator to calculate these multiplication problems.

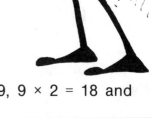

I'M SURE GLAD THEY DREW ME WITH TEN FINGERS.

1. 9 × 2 =   9 × 5 =   9 × 8 =

2. 9 × 1 =   9 × 7 =   9 × 3 =

3. 9 × 6 =   9 × 4 =   9 × 9 =

Bonus: Add the digits of each nine's product. Example: 9 × 1 = 9, 9 × 2 = 18 and 1 + 8 = 9, 9 × 3 = 27 and 2 + 7 = 9, etc.

GA1136

Decode and Multiply!

Example:

☐ × ⌐ = 5 × 9 = 45

Use the code to decode each multiplication problem before you find the product.

1. ⊏ × ⊐ =   ∟ × ☐ =   ∟ × ∟ =

2. ⌐ × ⊐ =   ⌐ × ⊐ =   ☐ × ⌟ =

3. ∟ × ⌐ =   ⊐ × ⊐ =   ⊔ × ∟ =

4. ⌐ × ⌐ =   ⊓ × ∟ =   ⊐ × ⊔ =

5. ⊓ × ⌐ =   ⌐ × ⌐ =   ⌟ × ⌟ =

6. ⌟ × ∟ =   ⊐ × ⌟ =   ⊏ × ⌟ =

7. ⊔ × ⌐ =   ⊓ × ⊔ =   ⊔ × ⊔ =

8. ⊏ × ⊔ =   ⊔ × ⌟ =   ⌐ × ⊔ =

Bonus: Leaving out any zeros, write your phone number in code. Multiply by two. Can you write your answer in code?

GA1136

Counting Cubes

Use multiplication facts to help you count the number of small cubes contained in each large cube below. Write each multiplication problem needed to solve the puzzles.

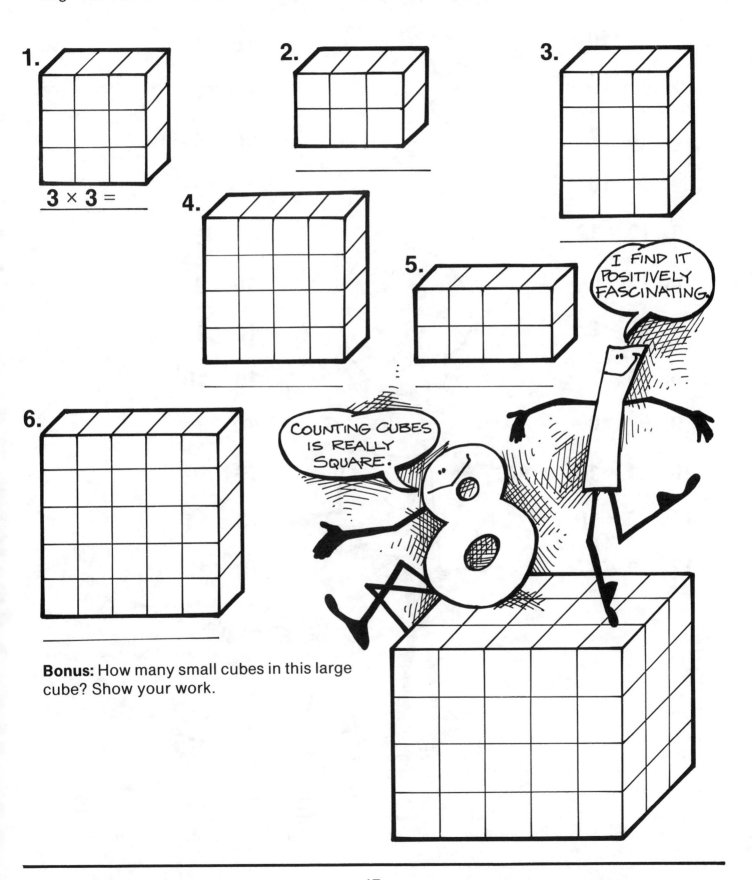

1.

$3 \times 3 =$ _____

2.

_____

3.

_____

4.

5.

I FIND IT POSITIVELY FASCINATING.

6.

COUNTING CUBES IS REALLY SQUARE.

Bonus: How many small cubes in this large cube? Show your work.

Multiplying by Ten

Multiplying a number by ten is simple. Just put a zero at the end of the number. Solve the multiplication problems below.

1. $3 \times 10 =$

2. $5 \times 10 =$

3. $10 \times 7 =$

4. $6 \times 10 =$

5. $10 \times 12 =$

6. $1 \times 10 =$

7. $10 \times 2 =$

8. $10 \times 8 =$

9. $9 \times 10 =$

10. $11 \times 10 =$

11. $12 \times 10 =$

12. $10 \times 4 =$

$10 \times 1 =$

$2 \times 10 =$

$10 \times 8 =$

$7 \times 10 =$

$10 \times 5 =$

$10 \times 6 =$

$10 \times 9 =$

$10 \times 10 =$

$4 \times 10 =$

$10 \times 3 =$

$10 \times 11 =$

$10 \times 12 =$

THE MOST POPULAR SHOE EVER IS NAMED AFTER ME!

NOT TO MENTION WE'RE THE BASE OF THE WHOLE NUMBER SYSTEM

Bonus: Can you write the product of one million, one hundred thousand times ten?

GA1136

Multiplying by Eleven

Continue numbering the chart below. Then use the chart to help you find the product for each multiplication problem.

1	2	3	4	5	6	7	8	9	10	11
12	13	14	15	16	17	18	19	20	21	22
23	24									
34	35									
45	46									
56										
67										
78										
89										
100										
111										
122										

HEY ELEVEN— CAN YOU DO ANYTHING CLEVER?

I CAN STAND ON MY HEAD AND STILL LOOK THE SAME

1. 11 × 1 =   11 × 2 =   11 × 3 =

2. 11 × 4 =   11 × 5 =   11 × 6 =

3. 11 × 7 =   11 × 8 =   11 × 9 =

4. 11 × 10 =   11 × 11 =   11 × 12 =

Bonus: Do you ever see a number pattern on the chart above? Describe an easy way to know the product of a number multiplied by 11.

Copyright © 1990, Good Apple, Inc.   19   GA1136

Multiplying by Twelve

A dozen is another way of saying twelve. Draw a dozen eggs in each carton below. Then use the egg cartons to help you solve the multiplication problems.

1. 1 × 12 =
2. 12 × 7 =
3. 12 × 6 =
4. 12 × 4 =
5. 12 × 8 =
6. 5 × 12 =
7. 10 × 12 =
8. 6 × 12 =

WHAT ABOUT A BAKER'S DOZEN?

3 × 12 =
8 × 12 =
12 × 5 =
12 × 1 =
12 × 2 =
9 × 12 =
12 × 12 =
12 × 11 =

7 × 12 =
2 × 12 =
4 × 12 =
12 × 3 =
12 × 9 =
12 × 10 =
11 × 12 =
12 × 0 =

Bonus: Which is greater, half a dozen times ten, or half of ten times a dozen?

GA1136

Multiplication Chart

Complete the multiplication chart. Some of the products have been written in for you.

✕	0	1	2	3	4	5	6	7	8	9	10	11	12
0	0	0	0	0	0								
1	0	1	2	3	4								
2	0	2	4	6									
3	0	3		9									
4	0	4			16								
5						25							
6							36						
7													
8													
9													
10													
11													
12													

Bonus: Color spaces containing an odd number PURPLE. Color spaces containing an even number YELLOW.

GA1136

Even and Odd Products

Find the product for each multiplication problem in the puzzle below. Then color spaces containing an even product ORANGE. Color spaces containing an odd product GREEN.

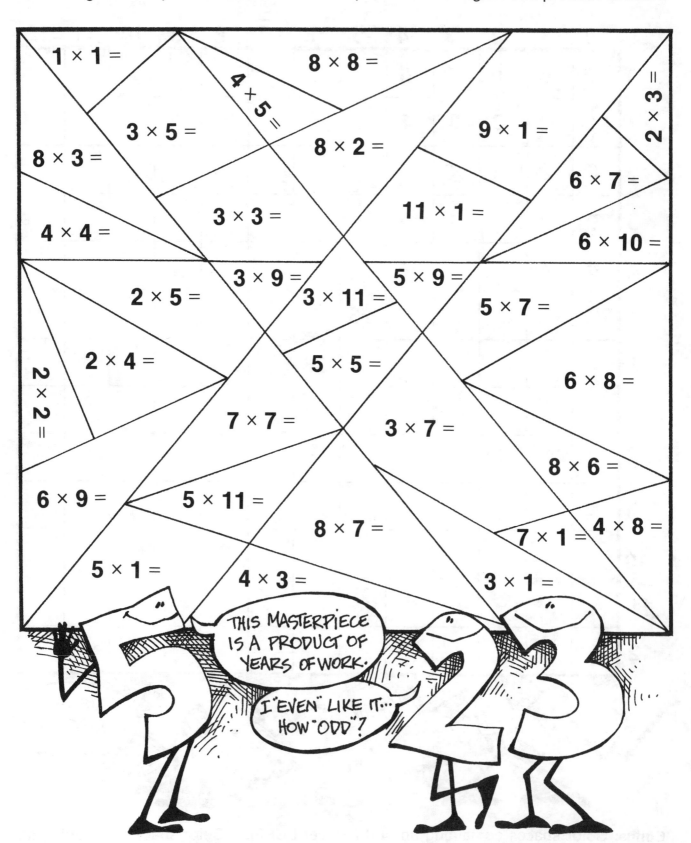

GA1136

Arrow Multiplication

Use the chart below to complete the multiplication problems. What looks like a secret code is actually just multiplication problems. The arrow tells you what direction to look on the chart to find the number to be multiplied by each number.
Example: 1 ➔ = 1 × 2. The arrow stands for the number at the right of 1.

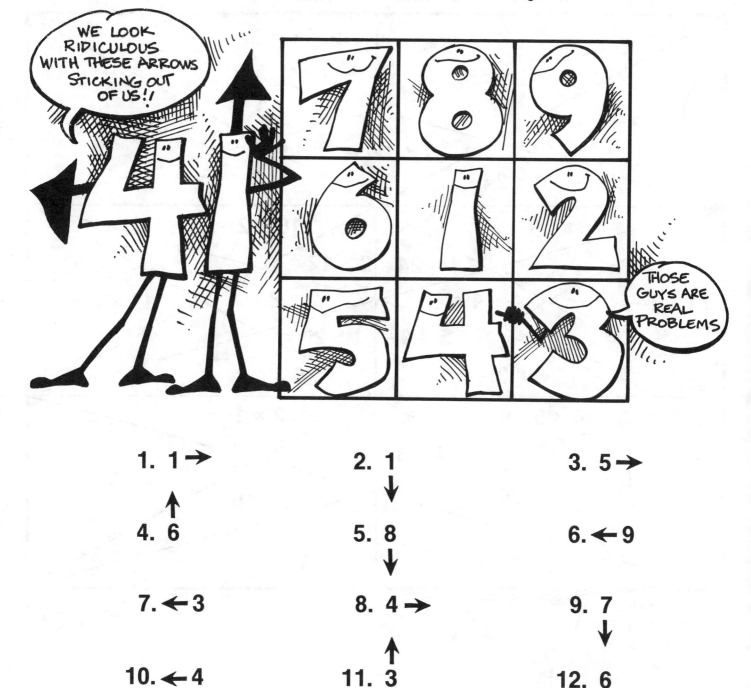

1. 1 ➔

2. 1 ↓

3. 5 ➔

4. 6 ↑

5. 8 ↓

6. ← 9

7. ← 3

8. 4 ➔

9. 7 ↓

10. ← 4

11. 3 ↑

12. 6 ↓

13. 7 ➔

14. 9 ↓

15. 2 ↓

Bonus: Write two problems in arrow code that have a product of 24. Write four problems in arrow code that have a product of 8.

GA1136

Multiplication Design

Complete each multiplication problem in the line design below.
Color sections of the design with a product of 12 ORANGE.
Color sections of the design with a product of 16 PINK.
Color sections of the design with a product of 20 RED.
Color sections of the design with a product of 24 BLUE.

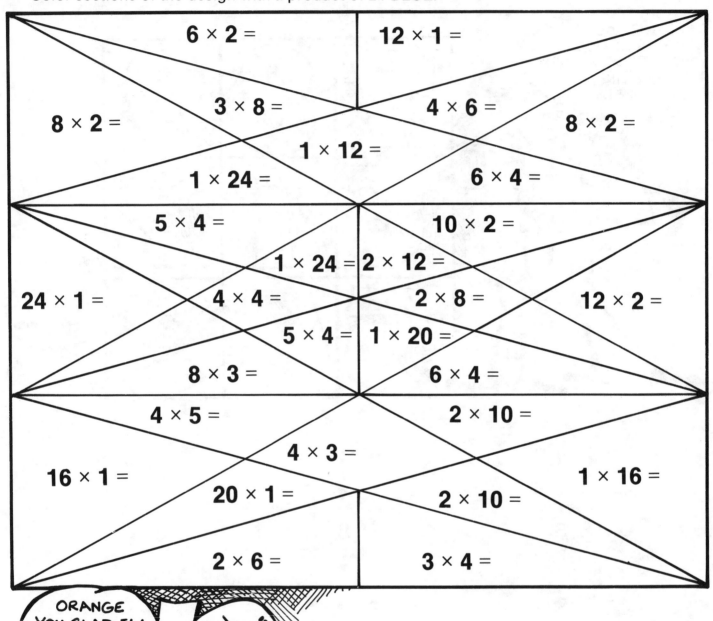

$6 \times 2 =$ $12 \times 1 =$

$8 \times 2 =$ $3 \times 8 =$ $4 \times 6 =$ $8 \times 2 =$

$1 \times 12 =$

$1 \times 24 =$ $6 \times 4 =$

$5 \times 4 =$ $10 \times 2 =$

$1 \times 24 =$ $2 \times 12 =$

$24 \times 1 =$ $4 \times 4 =$ $2 \times 8 =$ $12 \times 2 =$

$5 \times 4 =$ $1 \times 20 =$

$8 \times 3 =$ $6 \times 4 =$

$4 \times 5 =$ $2 \times 10 =$

$4 \times 3 =$

$16 \times 1 =$ $1 \times 16 =$

$20 \times 1 =$ $2 \times 10 =$

$2 \times 6 =$ $3 \times 4 =$

ORANGE YOU GLAD I'M AN ANSWER?

ORANGE

24

GA1136

Product Dot-to-Dot

Solve each multiplication problem below. Then connect each dot with a line, in the appropriate order, from problem one to seventeen.

1. **12 × 1 =**
2. **6 × 3 =**
3. **6 × 4 =**
4. **5 × 5 =**
5. **6 × 5 =**
6. **5 × 4 =**
7. **7 × 3 =**
8. **6 × 6 =**
9. **9 × 9 =**
10. **10 × 10 =**

11. **7 × 5 =**
12. **11 × 4 =**
13. **12 × 12 =**
14. **8 × 11 =**
15. **10 × 12 =**
16. **9 × 5 =**
17. **9 × 6 =**

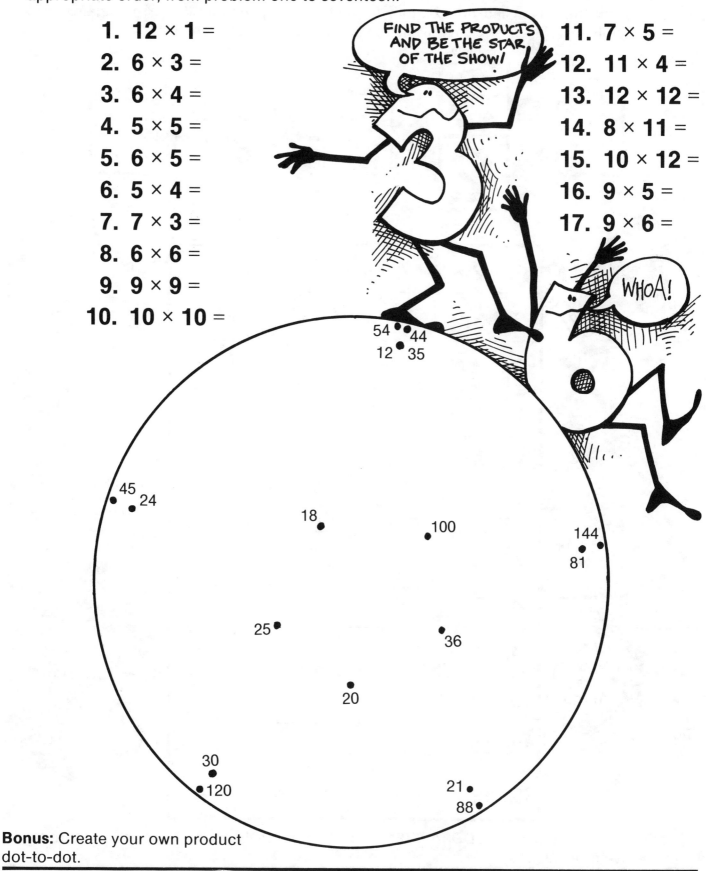

Bonus: Create your own product dot-to-dot.

GA1136

Computer Multiplication

Using each number below, complete the multiplication problems using the flow chart. Example: The first number is 2. Two is an even number, so we multiply by 2 ($2 \times 2 = 4$). Follow the arrow and add 2 more to 4 ($2 + 4 = 6$). Follow the arrow again, this time multiply by 2 ($6 \times 2 = 12$). Twelve is the answer to the first problem.

1. $2 \times 2 = 4 + 2 = 6 \times 2 = 12$

2. 3

3. 4

4. 5

5. 6

6. 7

7. 8

8. 9

9. 10

10. 11

11. 12

12. 13

CAN YOU SOLVE THIS?

ALL FOR NOTHING

Bonus: Change the word ODD to EVEN and EVEN to ODD on the flow chart and redo the twelve problems.

26

GA1136

Use the code below to write each multiplication problem. Then find the product for each problem.

1 = •	4 = •—	7 = ⬚	10 = ⬚
2 = • •	5 = •·•—	8 = ⬚	11 = ⬚
3 = • • •	6 = •·•·•—	9 = ⬚	12 = ⬚

1. ⬚ × ⬚ =

2. ⬚ × ⬚ =

3. ⬚ × ⬚ =

4. ⬚ × ⬚ =

5. ⬚ × ⬚ =

6. ⬚ × ⬚ =

7. ⬚ × ⬚ =

8. ⬚ × ⬚ =

9. ⬚ × ⬚ =

10. ⬚ × ⬚ =

11. ⬚ × ⬚ =

12. ⬚ × ⬚ =

Bonus: Multiply the year of your birth by the number of the month you were born. Example: If you were born in July 1981, your problem will be 1981 × 7 = •/•••/⬚/•••/⬚ Write the answer in code. (Jan. = 1, Feb. = 2, March = 3, etc.)

Multiplication Towers

Multiply the number pairs connected by a line and place the product in the circle at the top of each number pair. Work your way to the top of each tower. Some numbers have been filled in for you.

Bonus: Beginning with the seven digits in your phone number, build your own multiplication tower. How high can you go?

28

Search and Circle

Find and circle any true multiplication sentences in the number maze below. Number sentences may be hidden across, down or diagonally. Example: (5 × 6 = 30) so 5, 6 and 30 are circled in the top left-hand row. Can you find and circle thirty multiplication sentences? Forty? Fifty?

5 × 6 = 30			8	9	72	0	1	1	1	2
5	2	8	10	2	20	0	1	0	2	5
25	4	4	80	18	2	0	0	9	2	10
4	12	48	8	11	88	1	3	1	3	11
10	11	2	8	12	96	8	0	4	12	2
40	8	44	3	5	15	8	4	4	0	22
2	4	8	4	9	36	5	0	5	0	3
4	8	32	1	9	9	0	8	2	6	2
7	6	12	72	1	5	5	0	10	0	6
28	3	1	7	7	49	7	9	63	1	3
4	6	24	3	7	8	56	7	12	84	1
5	4	4	3	11	33	3	7	21	4	3
20	48	4	10	9	3	12	3	36	6	4
8	2	16	30	0	27	3	8	24	3	12

Bonus: Create your own search and circle multiplication maze. How many multiplication sentences can you hide? Give your puzzle to a friend to solve.

GA1136

What's Your Sign?

If you put the multiplication sign and equal sign in the right place in each group of digits below, you will make a true multiplication sentence. Example: 4 4 1 6 is 4 × 4 = 16 and 1 0 5 2 is 10 = 5 × 2

1.	8 2 1 6	6 1 1 6 6
2.	1 1 1 2 1 3 2	1 2 0 1 0 1 2
3.	2 7 1 4	1 1 0 1 0
4.	8 6 4 8	7 1 1 7 7
5.	7 1 0 7 0	1 2 1 1 2
6.	6 6 3 6	2 2 1 1 2
7.	1 2 6 2	1 1 1 1 1
8.	8 8 6 4	5 7 3 5
9.	2 0 1 0 2	5 8 4 0
10.	1 2 2 2 4	1 2 8 9 6
11.	3 6 1 8	8 0 1 0 8
12.	4 3 1 2	8 9 7 2
13.	1 0 5 5 0	5 9 4 5
14.	6 7 4 2	5 1 1 5 5
15.	6 9 5 4	5 1 2 6 0
16.	1 2 1 2 1 4 4	1 0 6 6 0

Bonus: Place the multiplication sign and equal sign in the correct place to make a true multiplication sentence. **1 2 0 0 0**

30

GA1136

More Multiplication Magic Squares

Complete each multiplication magic square. The first two numbers in each row and each column must have a product of the third number.

1.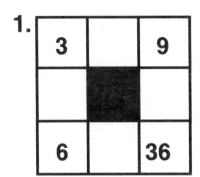

2.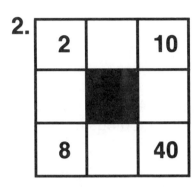

3.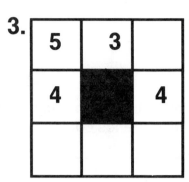

4.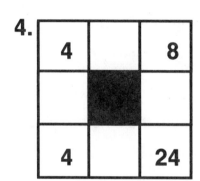

5.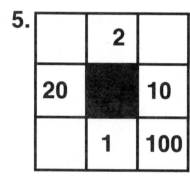

6.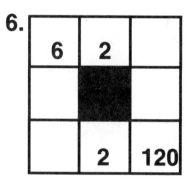

Bonus: Can you complete this multiplication magic square?

Perfectly
Preposterous Problems

Write a multiplication problem and product for each story problem below.

1. Seven silky snakes each had six slimy snails slithering on them. How many slimy snails slithered on the silky snakes?

_____

2. Four fabulous fountains each contained five freckled-faced friends. How many freckled-faced friends could be found in the fabulous fountains?

_____

3. Ten terrific turkeys each twirled three tiny tops. How many tiny tops did the ten terrific turkeys twirl?

_____

4. Nine nutty newts knitted no nightcaps. How many night caps did the nine nutty newts knit?

_____

5. Two timid turtles each took twelve tiny steps. How many tiny steps did the two timid turtles take?

_____

Bonus: If one ornery octopus had seven toes on each tentacle and wore three rings on each toe, how many rings did the ornery octopus wear?

_____

GA1136

Use the code to complete the multiplication problems. Example: $\langle$ = 3, $\langle\!\!\cdot$ = 7, $\langle\!\!\vdots$ = 11

1. $\vee\!\!\cdot$ × $\wedge\!\!\cdots$ = **120** $\wedge\!\!\cdots$ × $\vee\!\!\cdot$ = $\vdots\rangle$ × $\vdots\rangle$ =

2. $\wedge\!\!\cdot$ × $\vdots\rangle$ = $\vee\!\!\cdot$ × $\vee\!\!\cdot$ = $\vee\!\!\cdot$ × $\vee\!\!\cdot$ =

3. $\wedge\!\!\cdot$ × $\wedge\!\!\cdot$ = $\wedge\!\!\cdots$ × $\cdot\rangle$ = $\wedge\!\!\cdot$ × $\wedge\!\!\cdots$ =

4. $\langle\!\!\cdot$ × $\langle\!\!\cdot$ = $\langle\!\!\cdot$ × $\wedge\!\!\cdots$ = $\cdot\rangle$ × $\wedge\!\!\cdots$ =

5. $\wedge\!\!\cdot$ × $\vee\!\!\cdot$ = $\langle\!\!\cdot$ × $\vdots\rangle$ = $\vee\!\!\cdot$ × $\vdots\rangle$ =

6. $\vdots\rangle$ × $\langle\!\!\vdots$ = $\langle\!\!\cdot$ × $\cdot\rangle$ = $\wedge\!\!\cdot$ × $\langle\!\!\vdots$ =

7. $\vdots\rangle$ × $\vee\!\!\cdot$ = $\wedge\!\!\cdots$ × $\wedge\!\!\cdots$ = $\wedge\!\!\cdots$ × $\wedge$ =

8. $\langle\!\!\cdot$ × $\vee\!\!\cdot$ = $\langle\!\!\vdots$ × $\vee\!\!\cdot$ = $\langle\!\!\vdots$ × $\wedge\!\!\cdots$ =

Bonus: Write six different problems that have the same product using the "X" code.

GA1136

Multiplication Connection

Draw a line to connect number pairs with the appropriate product. The first one has been completed for you.

1	12 = 72
2	11 = 11
3	10 = 70
4	9 = 45
5	8 = 64
6	7 = 28
7	6 = 18
8	5 = 45
9	4 = 40
10	3 = 6
11	2 = 22
12	1 = 12

1	12 = 24
2	11 = 44
3	10 = 80
4	9 = 27
5	8 = 72
6	7 = 7
7	6 = 36
8	5 = 35
9	4 = 20
10	3 = 36
11	2 = 20
12	1 = 11

THESE PROBLEMS HAVE ME ALL TIED UP!

Bonus: Make up your own multiplication connection puzzle.

GA1136

Secret Message

To discover the Secret Message, work the multiplication problem in each square below. Then place the letters in the appropriate blanks.

A $\begin{array}{r} 3 \\ \times\,6 \\ \hline \end{array}$	F $\begin{array}{r} 7 \\ \times\,7 \\ \hline \end{array}$	O $\begin{array}{r} 8 \\ \times\,7 \\ \hline \end{array}$	Y $\begin{array}{r} 0 \\ \times\,2 \\ \hline \end{array}$	P $\begin{array}{r} 5 \\ \times\,5 \\ \hline \end{array}$
G $\begin{array}{r} 11 \\ \times\,8 \\ \hline \end{array}$	B $\begin{array}{r} 3 \\ \times\,1 \\ \hline \end{array}$	Q $\begin{array}{r} 2 \\ \times\,1 \\ \hline \end{array}$	N $\begin{array}{r} 12 \\ \times\,8 \\ \hline \end{array}$	R $\begin{array}{r} 6 \\ \times\,5 \\ \hline \end{array}$
S $\begin{array}{r} 6 \\ \times\,6 \\ \hline \end{array}$	T $\begin{array}{r} 11 \\ \times\,12 \\ \hline \end{array}$	C $\begin{array}{r} 10 \\ \times\,1 \\ \hline \end{array}$	H $\begin{array}{r} 8 \\ \times\,9 \\ \hline \end{array}$	M $\begin{array}{r} 3 \\ \times\,3 \\ \hline \end{array}$
V $\begin{array}{r} 9 \\ \times\,9 \\ \hline \end{array}$	U $\begin{array}{r} 12 \\ \times\,7 \\ \hline \end{array}$	I $\begin{array}{r} 4 \\ \times\,4 \\ \hline \end{array}$	D $\begin{array}{r} 10 \\ \times\,9 \\ \hline \end{array}$	L $\begin{array}{r} 10 \\ \times\,10 \\ \hline \end{array}$
K $\begin{array}{r} 8 \\ \times\,8 \\ \hline \end{array}$	J $\begin{array}{r} 2 \\ \times\,2 \\ \hline \end{array}$	W $\begin{array}{r} 2 \\ \times\,3 \\ \hline \end{array}$	X $\begin{array}{r} 1 \\ \times\,1 \\ \hline \end{array}$	E $\begin{array}{r} 3 \\ \times\,4 \\ \hline \end{array}$

Secret Message:

__ __ __   __ __ __   __ __ __ __   __ __ __ __ __ __ __
0 56 84   18 30 12   81 12 30 0   36 25 12 10 16 18 100

Bonus: If Z = 17, write your full name in number code. Write your telephone number in letter code.

35

Colored Spots

Complete the multiplication problem in each spot. Color spots with a product of 12 ORANGE. Color spots with a product of 20 YELLOW. Color spots with a product of 24 PURPLE. Color spots with a product of 30 GREEN.

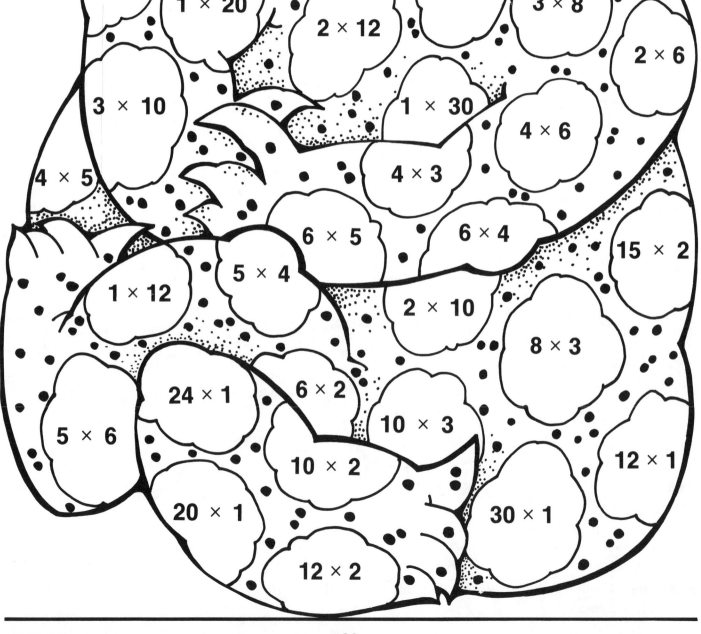

MULTIPLICATION CAN BE MONSTROUS.

36

Multiplication Crossword

Across:

1. $3 \times 7 =$
2. $10 \times 2 =$
3. $5 \times 6 =$
4. $12 \times 2 =$
5. $9 \times 10 =$
6. $5 \times 8 =$
7. $10 \times 3 =$
8. $10 \times 8 =$
9. $7 \times 6 =$
10. $3 \times 12 =$
11. $1 \times 1 =$

12. $10 \times 9 =$
13. $8 \times 11 =$
14. $12 \times 8 =$
15. $3 \times 12 =$
16. $3 \times 8 =$
17. $7 \times 7 =$
18. $1 \times 12 =$

Down:

1. $4 \times 5 =$
2. $10 \times 2 =$
3. $10 \times 3 =$

4. $5 \times 4 =$
5. $10 \times 9 =$
6. $6 \times 7 =$
7. $9 \times 4 =$
8. $8 \times 10 =$
9. $8 \times 6 =$
10. $12 \times 3 =$
11. $2 \times 8 =$
12. $8 \times 12 =$
13. $7 \times 12 =$
14. $9 \times 11 =$
15. $8 \times 4 =$

Bonus: Create your own multiplication crossword puzzle.

 GA1136

Multiplication Paths

Begin at the top number and trace each path, multiplying the two numbers in each circle. Write the product of each number pair in the circle at the end of the path.

Bonus: Create your own multiplication paths maze.

GA1136

Multiply the number in the center of each bull's-eye by each of the numbers in the inner ring. Write the product in the outside ring. The first one has been started for you.

Products Quilt

Complete the multiplication graph. Then color the spaces with the same number the same color. Color the number squares in the first column and along the top row in matching colors also. All boxes containing the digit 1 will be colored the same color. All boxes containing the digit 2 will be colored the same color, but a different color than boxes containing the digit 1.

Bonus: What products are found only once in the graph?

Multiplication Boggler

Arrange the digits 1, 2, 3, 4 and 6 in the circles so that each of the four lines of three circles is a true multiplication sentence. The product is in the center circle.

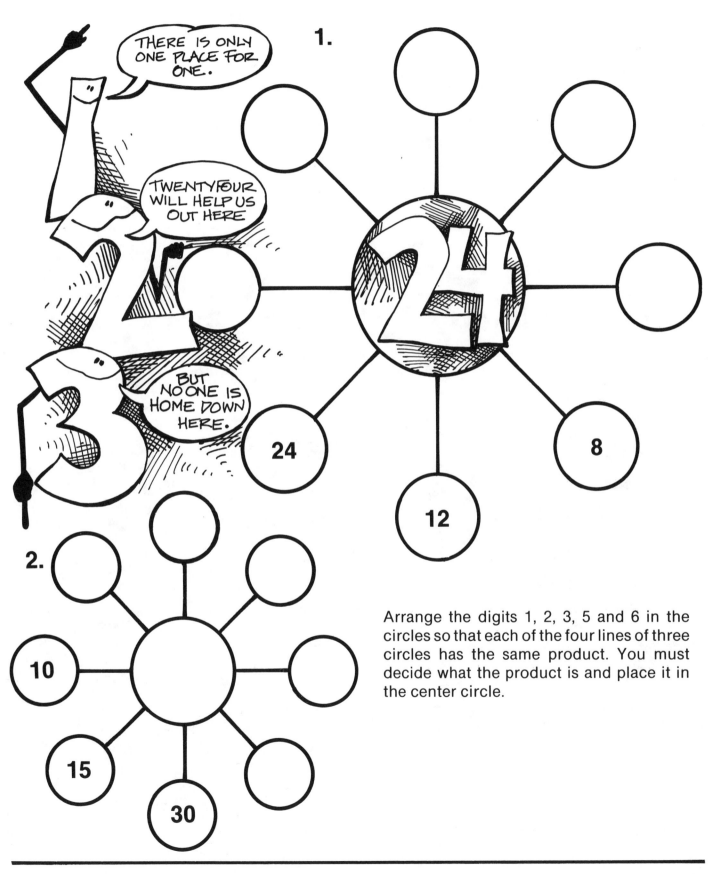

1.

2.

Arrange the digits 1, 2, 3, 5 and 6 in the circles so that each of the four lines of three circles has the same product. You must decide what the product is and place it in the center circle.

__Full Circle Multiplication__

Hidden in each circle are enough numbers to make a true multiplication sentence. Can you use all the digits in each circle to create a multiplication sentence? The first one has been completed for you.

1.

12 × 3 = 36

2.

3.

4.

Bonus: Can you write a true multiplication sentence using these digits: 0 0 0 1 2?

Circle Three

Find and circle ten true multiplication sentences in each box below. Every digit will be circled one time only.

Box 1:

(3 × 6 = 18) (9 × 2 = 18)

5	9	1	9	11	12
2	2	2	4	1	×1
10	7	2	14	11	12
2	3	6	2	1	2
4	2	8	6	2	12

Box 2:

10	2	8	16	9
2	3	3	9	9
20	3	4	12	81
2	11	22	3	9
2	12	24	5	10
3	2	6	15	90

Box 3:

9	11	99	9	12	108
11	1	1	1	2	3
0	10	1	10	1	1
0	5	1	5	2	3
12	0	0	4	1	4
10	8	80	10	0	0

Box 4:

6	1	6	7	2	0
3	3	9	1	11	6
2	12	11	7	22	0
3	2	2	10	0	0
6	24	22	9	0	0
5	7	35	6	5	30

THE APPROPRIATE PLACEMENT OF MY EXTREMELY ATTRACTIVE SELF, IN CONJUNCTION WITH A SIGN OF EQUALITY, WILL PRODUCE A TRUE MULTIPLICATION SENTENCE OF THE UTMOST SIGNIFICANCE.

IT SAID A MOUTHFUL!

43

GA1136

Product Designs

Write number pairs in the circles connected by a line that have a product of the number in the center of each design. The first one has been completed for you. Example: 6 × 1 = 6 and 3 × 2 = 6, so the number pairs 6, 1 and 3, 2 have been placed in the first design.

Bonus: Draw a number design for the product 36.

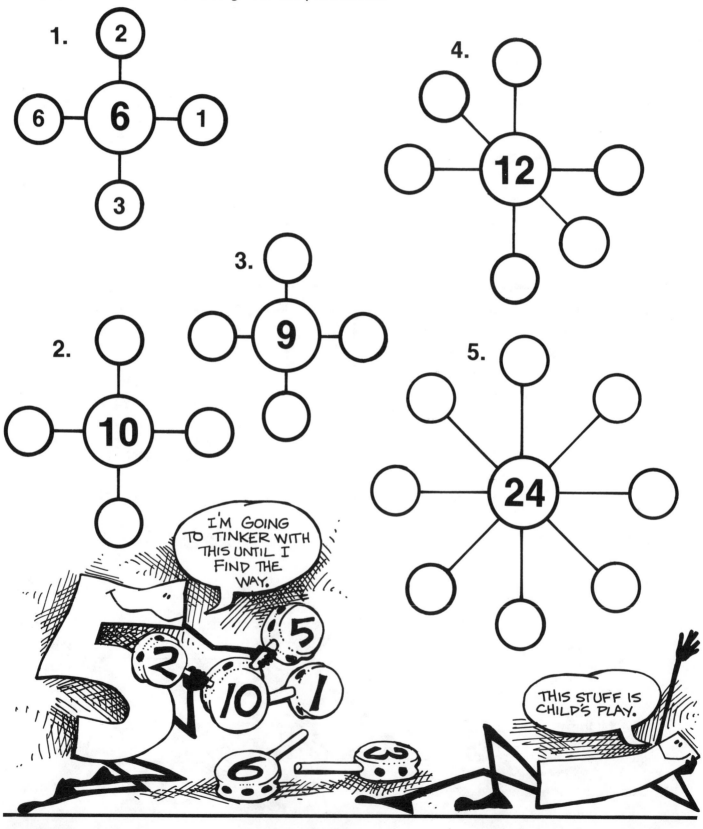

GA1136

Product Pairs

Circle number pairs in each box with a product of the number in the left-hand corner. Don't circle numbers that do not have a product of the number in the corner. Example: The number pair 4 and 5 were circled in the first box because 4 × 5 = 20.

1. 20

2   4   20

3   7

1   5   10

2. 36

6   2

6   9   18

6 13   3

3   36

4

12 1

3. 12   6   12

5

2

2 3   4 1

4. 100 100 25

50   2

10   5

1   4

20 10

5. 120 60

5   30

12   10

4   24

2

6. 40   10

2   20

8   5

4

Bonus: What is the smallest pair of numbers with an odd product?

    GA1136

Product Matchup

To discover the Mystery Word, complete each multiplication problem below. Then draw a line connecting each multiplication problem in the left-hand column with a multiplication problem in the right-hand column that has the same product. Write the letters that are intersected by a line in the order they are found from left to right and top to bottom.

Mystery Word:

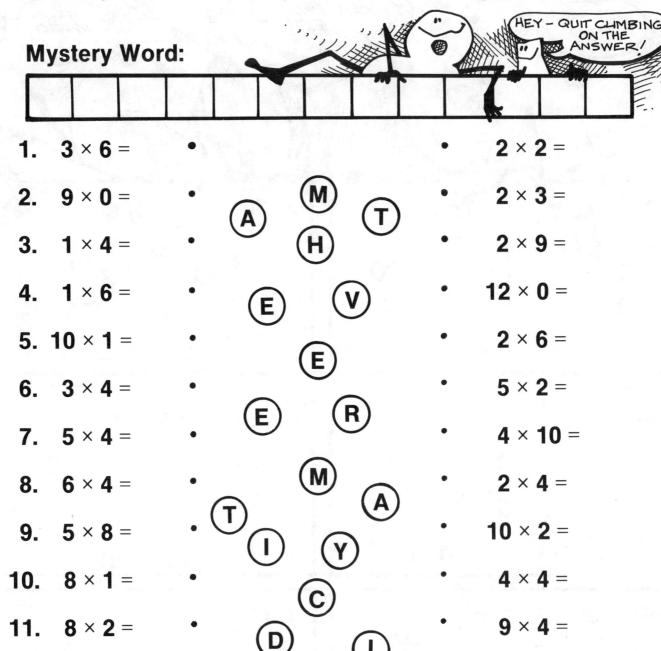

1. $3 \times 6 =$ •   • $2 \times 2 =$

2. $9 \times 0 =$ •   • $2 \times 3 =$

3. $1 \times 4 =$ •   • $2 \times 9 =$

4. $1 \times 6 =$ •   • $12 \times 0 =$

5. $10 \times 1 =$ •   • $2 \times 6 =$

6. $3 \times 4 =$ •   • $5 \times 2 =$

7. $5 \times 4 =$ •   • $4 \times 10 =$

8. $6 \times 4 =$ •   • $2 \times 4 =$

9. $5 \times 8 =$ •   • $10 \times 2 =$

10. $8 \times 1 =$ •   • $4 \times 4 =$

11. $8 \times 2 =$ •   • $9 \times 4 =$

12. $6 \times 8 =$ •   • $12 \times 4 =$

13. $6 \times 6 =$ •   • $8 \times 3 =$

14. $5 \times 6 =$ •   • $10 \times 3 =$

Bonus: Unscramble the unused letters to find out the best day to study multiplication facts.

GA1136

Hidden Number Pairs

Hidden in each row of digits below are number pairs with a product of the last number in that row. Find and circle the number pairs. For example, the number pairs 2, 5 and 10, 1 are circled in the top row because 2 × 5 = 10 and 10 × 1 = 10. Can you find yet another pair in the top row with a product of 10?

1.	(2	5)	8	6	(10	1)	3	(2	5)	= 10
2.	6	6	9	4	12	2	18	36	1	= 36
3.	5	4	7	3	2	10	4	5	8	= 20
4.	3	6	9	2	18	1	16	2	6	= 18
5.	16	1	4	5	4	4	8	2	7	= 16
6.	5	3	3	4	12	1	6	2	5	= 12
7.	1	6	2	3	2	5	1	4	2	= 6
8.	3	4	5	3	15	1	15	2	7	= 15
9.	2	10	12	2	6	4	8	3	8	= 24
10.	8	4	16	2	32	1	20	12	8	= 32

ORANGE YOU A NUMBER PAIR?

PEAR-HAPS NOT.

BANEENER AM I.

Bonus: Circle three number pairs in the row below that have the same product.

0   1   6   2   4   3   0   7   1   9   2   6

47

Colorful Products

After you find the product for each problem below, color the boxes with the appropriate colors.

Products of 0-10 color RED.
Products of 11-20 color BLUE.
Products of 21-30 color YELLOW.
Products of 31-40 color PURPLE.
Products of 41-50 color GREEN.
Products of 50 or more color ORANGE.

10 × 1 =	12 × 3 =	4 × 8 =	11 × 4 =	11 × 1 =	3 × 5 =
10 × 5 =	8 × 4 =	11 × 3 =	9 × 1 =	4 × 11 =	
5 × 2 =	11 × 5 =	12 × 5 =	12 × 1 =		3 × 6 =
6 × 9 =	7 × 9 =	6 × 0 =	12 × 4 =	10 × 4 =	9 × 6 =
1 × 1 =	10 × 6 =	FACTS ARE THE BUILDING BLOCKS OF MATHEMATICS.	6 × 2 =	4 × 4 =	
10 × 7 =	8 × 0 =		8 × 7 =	4 × 10 =	
4 × 9 =	10 × 10 =		7 × 8 =	2 × 1 =	
7 × 2 =	4 × 5 =		10 × 0 =	6 × 10 =	
9 × 11 =			9 × 4 =	7 × 3 =	5 × 5 =
8 × 3 =	4 × 6 =		3 × 1 =	3 × 3 =	
9 × 3 =	10 × 3 =		6 × 4 =	4 × 7 =	
12 × 2 =	2 × 11 =		3 × 7 =	5 × 6 =	
3 × 2 =	4 × 1 =	2 × 12 =	11 × 2 =	3 × 8 =	7 × 4 =
7 × 5 =	11 × 6 =	8 × 9 =	2 × 5 =	8 × 2 =	2 × 6 =
5 × 1 =	6 × 11 =	11 × 7 =	5 × 7 =	8 × 5 =	9 × 9 =
2 × 4 =	8 × 8 =	9 × 2 =	2 × 7 =	9 × 7 =	
6 × 1 =	12 × 7 =	5 × 8 =	9 × 5 =	2 × 3 =	9 × 10 =
10 × 8 =	10 × 2 =	2 × 9 =	11 × 8 =	12 × 8 =	
7 × 1 =	4 × 12 =	2 × 2 =	8 × 4 =	10 × 4 =	5 × 9 =
3 × 4 =	4 × 3 =	5 × 10 =	3 × 11 =	6 × 6 =	8 × 1 =

Consecutive Products

Consecutive numbers are numbers that follow in counting order. Example: 6 and 7 are consecutive because seven always follows six when counting. Use consecutive number pairs to solve each problem below.

1. What two consecutive numbers have a product of 6? _____

2. What two consecutive numbers have a product of 20? _____

3. What two consecutive numbers have a product of 0? _____

4. What two consecutive numbers have a product of 12? _____

5. What two consecutive numbers have a product of 90? _____

6. What two consecutive numbers have a product of 30? _____

7. What two consecutive numbers have a product of 110? _____

8. What two consecutive numbers have a product of 72? _____

9. What two consecutive numbers have a product of 56? _____

10. What two consecutive numbers have a product of 156? _____

Bonus: What two consecutive number pairs have a larger total than product?

GA1136

Multiplication Patterns

Find the pattern in each row of numbers below. Continue each row of numbers to include three more numbers. Then write the number pattern on the line after each row. Example: 2 3 6 7 14 (The pattern is add one, multiply by two, add one, multiply by two, repeat, etc.) So the next three numbers will be 15, 30, 31.

1. | 8 | 10 | 10 | 12 | 12 | | | | _____

2. | 1 | 1 | 2 | 2 | 4 | | | | _____

3. | 2 | 4 | 8 | 10 | 20 | | | | _____

4. | 1 | 11 | 22 | 32 | 64 | | | | _____

5. | 20 | 10 | 20 | 10 | 20 | | | | _____

6. | 1 | 1 | 2 | 6 | 24 | | | | _____

Bonus: Can you extend this letter pattern?   What is the pattern?

| O | T | T | F | F | S | S | E | N | | _____

50

Get This Straight

Draw two straight lines to divide the square so that each area has two numbers with a product of 12.

Draw two straight lines to divide the square so that each area has two numbers with a product of 20.

Draw two straight lines to divide the square so that each area has two numbers with a product of 24.

Draw two straight lines to divide the square so that each area has two numbers with a product of 36.

51

Three Straight Lines

Draw three straight lines to divide the box. Each area the lines create should contain two numbers whose product is 36.

Draw three straight lines to divide the box. Each area the lines create should contain two numbers whose product is 24.

Draw three straight lines to divide the box. Each area the lines create should contain two numbers whose product is 12.

Draw three straight lines to divide the box. Each area the lines create should contain two numbers whose product is 40.

Bonus: Create your own box puzzle for the following products: 50, 100 and 120.

GA1136

Making Arrangements

Each multiplication problem below contains the four digits 1, 2, 3 and 4. Each answer is given. You must place the correct missing digits in each problem. Example: The digits 1 and 3 are missing on the first problem. Which is correct 421 × 3 = 423 or 423 × 1 = 423?

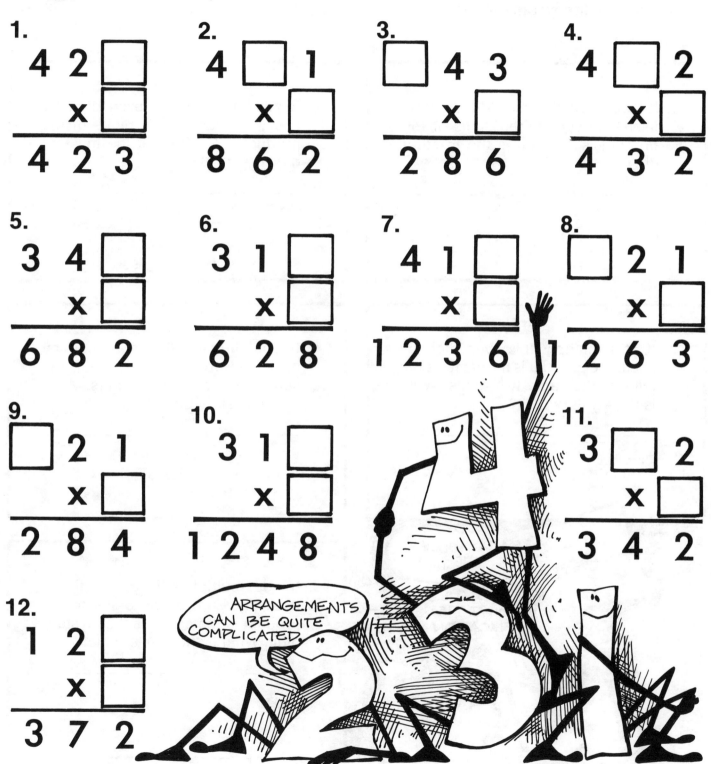

1.
4 2 ☐
× ☐
___
4 2 3

2.
4 ☐ 1
× ☐
___
8 6 2

3.
☐ 4 3
× ☐
___
2 8 6

4.
4 ☐ 2
× ☐
___
4 3 2

5.
3 4 ☐
× ☐
___
6 8 2

6.
3 1 ☐
× ☐
___
6 2 8

7.
4 1 ☐
× ☐
___
1 2 3 6

8.
☐ 2 1
× ☐
___
1 2 6 3

9.
☐ 2 1
× ☐
___
1 2 8 4

10.
3 1 ☐
× ☐
___
1 2 4 8

11.
3 ☐ 2
× ☐
___
3 4 2

12.
1 2 ☐
× ☐
___
3 7 2

Bonus: Make up your own missing digits puzzle using the odd digits 1, 3, 5 and 7.

GA1136

Mystery Digits

Each problem below contains hints for finding a mystery digit. For example, the first problem calls for an odd digit. You know that the odd digits are 1, 3, 5, 7 and 9. Which of those numbers when multiplied by itself has a product with digits that total 7? Try each one to discover the answer.

1. If you multiply this odd digit by itself, the sum of the digits of the product is 7. What is the mystery digit?

_____

2. If you multiply this even digit by itself, the sum of the digits of the product is 9. What is the mystery digit?

_____

3. If you multiply this even digit by itself, the product is the same as the sum of the same two digits. What is the mystery digit?

_____

4. If you multiply this odd digit by itself, the product is less than the sum of the same two digits. What is the mystery digit?

_____

DO YOU HAVE THE TIME?

IT'S TIME TO SOLVE THESE MULTI-PLICATION MYSTERIES.

5. If you multiply this odd digit by itself, the sum of the digits of the product is the same as the mystery digit. What is the mystery digit?

_____

Bonus: If you multiply this even digit by itself, the product is the same as the sum of the two same digits. What is the mystery digit?

                    GA1136

Multiplication Magic

1. Write any one-digit number in the first box.
2. Multiply the number you have chosen by 9.
3. Multiply 12345679 by the product found in step 2.
4. Repeat steps 1-3 choosing a different digit each time in the other three boxes.

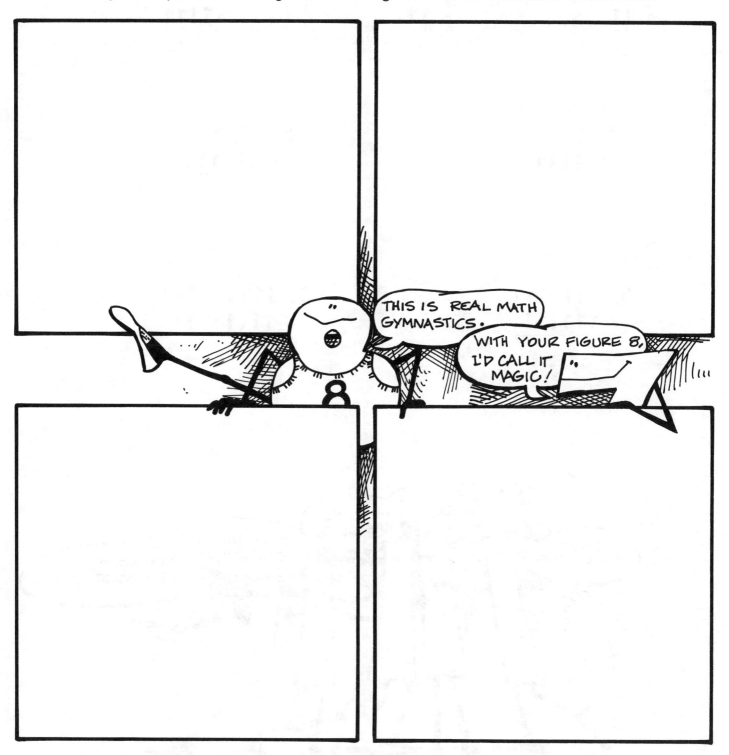

Bonus: Choose another digit. Multiply it by 9. Then multiply 123456789 by the product. Repeat three more times. What new pattern do you see?

GA1136

Magical Eleven

Complete each multiplication problem below.

1. 11
    × 11
   ____

2. 111
    × 111
   _____

3. 1111
    ×1111
   _____

4. 11111
    × 11111
   _____

5. 111111
    × 111111
   _____

6. 1111111
    × 1111111
   _____

7. 11111111
    × 11111111
   _____

Bonus: What is 111,111,111 × 111,111,111?

GA1136

Dropping Zeros

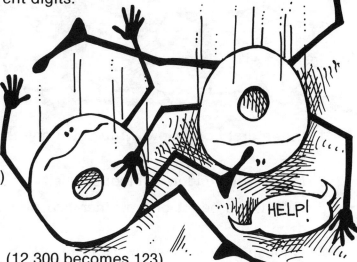

1. Choose a number that contains three different digits. (123)

2. Multiply the number by 2. (123 × 2 = 246)

3. Add 4 to the product. (246 + 4 = 250)

4. Multiply the sum by 5. (250 × 5 = 1250)

5. Add 12 to the product. (1250 + 12 = 1262)

6. Multiply the sum by 10. (1262 × 10 = 12,620)

7. Subtract 320 from the product. (12,620 − 320 = 12,300)

8. Drop all the zeros at the end of the answer. (12,300 becomes 123)

9. Repeat steps 1-8 five more times, once in each box, choosing different three-digit numbers.

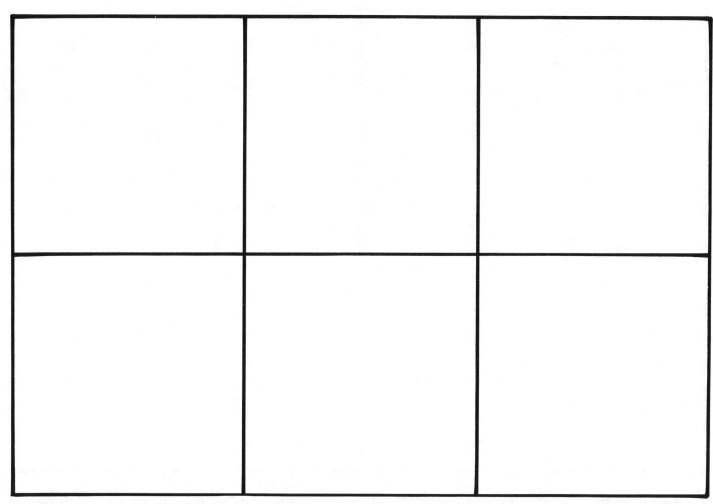

Bonus: What pattern did you discover?

GA1136

Pick Three, Any Three

1. In the first box below, write down any three numbers listed consecutively in the row above.

2. Multiply the middle number by itself.

3. Find the product of the other two numbers. Compare your answer in steps 2 and 3.

4. Choosing three new consecutively listed numbers each time, repeat steps 1-3 in each of the other three boxes.

Bonus: What pattern do you see in the row of numbers? Extend the row to include three more numbers?

GA1136

Calculator Magic

Use a calculator to help you find the product for each problem below.

1. $9 \times 9 + 7 =$
2. $9 \times 98 + 6 =$
3. $9 \times 987 + 5 =$
4. $9 \times 9876 + 4 =$
5. $9 \times 98765 + 3 =$
6. $9 \times 987654 + 2 =$
7. $9 \times 9876543 + 1 =$
8. $9 \times 98765432 + 0 =$

Bonus: Using a calculator, complete the problems below.

1. $1 \times 9 + 2 =$
2. $12 \times 9 + 3 =$
3. $123 \times 9 + 4 =$
4. $1234 \times 9 + 5 =$
5. $12345 \times 9 + 6 =$
6. $123456 \times 9 + 7 =$
7. $1234567 \times 9 + 8 =$
8. $12345678 \times 9 + 9 =$

GA1136

2 × 1 =	**10 × 8** =	**7 × 0** =
2 × 3 =	**8 × 6** =	**7 × 9** =
3 × 5 =	**7 × 1** =	**11 × 5** =
2 × 0 =	**4 × 6** =	**8 × 7** =
7 × 4 =	**3 × 1** =	**4 × 2** =
11 × 0 =	**12 × 4** =	**10 × 7** =
10 × 2 =	**10 × 5** =	**11 × 3** =
10 × 1 =	**3 × 0** =	**11 × 8** =
3 × 10 =	**11 × 7** =	**12 × 10** =
9 × 9 =	**11 × 12** =	**12 × 12** =

GA1136

$3 \times 3 =$	$8 \times 0 =$	$8 \times 1 =$
$12 \times 9 =$	$10 \times 4 =$	$7 \times 2 =$
$10 \times 10 =$	$4 \times 9 =$	$8 \times 3 =$
$2 \times 6 =$	$7 \times 6 =$	$4 \times 4 =$
$11 \times 9 =$	$12 \times 2 =$	$5 \times 6 =$
$12 \times 8 =$	$3 \times 7 =$	$8 \times 4 =$
$12 \times 11 =$	$7 \times 5 =$	$4 \times 0 =$
$9 \times 0 =$	$5 \times 1 =$	$11 \times 6 =$
$10 \times 6 =$	$9 \times 6 =$	$12 \times 1 =$
$8 \times 8 =$	$12 \times 7 =$	$12 \times 5 =$

GA1136

1 × 1 =	2 × 2 =	9 × 2 =
11 × 11 =	6 × 6 =	10 × 9 =
4 × 3 =	9 × 8 =	6 × 1 =
11 × 4 =	2 × 8 =	5 × 4 =
1 × 0 =	6 × 3 =	10 × 11 =
5 × 5 =	5 × 2 =	6 × 0 =
10 × 0 =	12 × 6 =	7 × 7 =
4 × 1 =	9 × 1 =	9 × 3 =
11 × 2 =	5 × 9 =	11 × 1 =
5 × 0 =	12 × 3 =	12 × 0 =

GA1136

Graphing Your Progress

Number Correct	1	2	3	4	5	6	7	8	9	10	11	12	13
90													
88													
86													
84													
82													
80													
78													
76													
74													
72													
70													
68													
66													
64													
62													
60													
58													
56													
54													
52													
50													
48													
46													
44													
42													
40													
38													
36													
34													
32													
30													
28													
26													
24													
22													
20													
18													
16													
14													
12													
10													
8													
6													
4													
2													
Test #	1	2	3	4	5	6	7	8	9	10	11	12	13

GA1136

Digit Races

The word *digit* comes from the Latin word meaning "fingers." Human fingers were truly the first "digital computers."

Object: To practice the multiplication facts 1 × 1 through 10 × 10

Rules: Two players may play this game. They stand facing each other with their hands behind their backs. They count together: "one, two and three." On the count of "three," they both extend one to ten fingers in front of them. Both players then calculate the product of the two numbers indicated by their fingers. The first person to shout out the correct answer is the winner. If a player shouts out an incorrect product, the other player gets the point and the players start a new round. Each player should play twenty rounds with one opponent and then change partners and play twenty rounds with the new opponent. Repeat with five different opponents. Use the handy score sheet found below. Simply put the initial of the winner of each round in the appropriate space.

Game		#1	#2	#3	#4	#5
Name of Players	1. 2.					
Round 1						
Round 2						
Round 3						
Round 4						
Round 5						
Round 6						
Round 7						
Round 8						
Round 9						
Round 10						
Round 11						
Round 12						
Round 13						
Round 14						
Round 15						
Round 16						
Round 17						
Round 18						
Round 19						
Round 20						
Winner						

GA1136

Twenty Rolls

Object: To practice the multiplication facts 1 × 1 through 6 × 6

Rules: Roll the dice twenty times. Each time you roll, record the product of the two numbers indicated on the dice on the score sheet along with the problem. After twenty products are recorded, the total game score is found by adding the twenty products. If you score two hundred and fifty points or more, you win! If you score less than two hundred and fifty points, you lose.

Roll #	Problem	Product
1		
2		
3		
4		
5		
6		
7		
8		
9		
10		
11		
12		
13		
14		
15		
16		
17		
18		
19		
20		
	Total	

HELP, IT'S RAINING DICE!

KEEP YOUR UMBRELLA HANDY AND MULTIPLY LIKE MAD!

GA1136

Products in a Flash

Object: To practice multiplication facts 0 × 12 through 12 × 12

Rules: To play this game you will need flash cards (see pages 67 and 68), a die, and a marker for each player. Players are to take turns turning over the top card and giving the product within five seconds. (Other players can slowly count: 1001, 1002, 1003, 1004, 1005.) If player gives the correct product, he/she rolls the die and advances marker appropriate number of spaces on the board. The first player to reach the end of the maze is declared the winner.

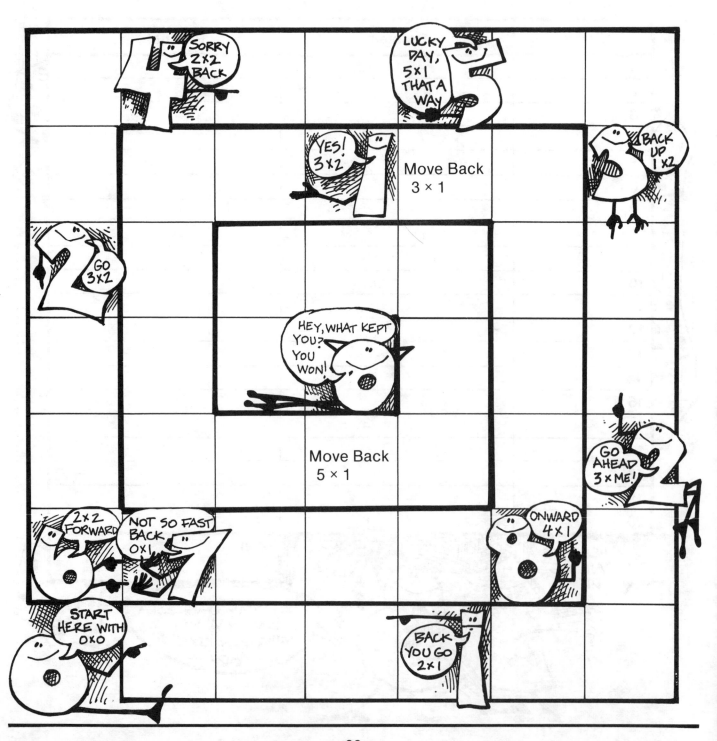

GA1136

$12 \times 12 =$	$11 \times 3 =$	$9 \times 9 =$
$8 \times 5 =$	$2 \times 11 =$	$7 \times 2 =$
$12 \times 6 =$	$9 \times 3 =$	$7 \times 5 =$
$6 \times 6 =$	$11 \times 12 =$	$11 \times 8 =$
$12 \times 5 =$	$4 \times 5 =$	$5 \times 5 =$
$6 \times 4 =$	$7 \times 7 =$	$11 \times 6 =$
$7 \times 3 =$	$9 \times 10 =$	$6 \times 3 =$
$10 \times 12 =$	$4 \times 4 =$	$11 \times 11 =$
$12 \times 8 =$	$6 \times 5 =$	$8 \times 2 =$
$5 \times 3 =$	$12 \times 0 =$	$6 \times 2 =$

GA1136

10 × 11 =	**12 × 4 =**	**9 × 11 =**
8 × 9 =	**12 × 3 =**	**7 × 4 =**
9 × 4 =	**10 × 3 =**	**11 × 4 =**
8 × 4 =	**2 × 2 =**	**5 × 2 =**
8 × 3 =	**7 × 8 =**	**11 × 2 =**
3 × 4 =	**10 × 10 =**	**8 × 8 =**
9 × 12 =	**12 × 7 =**	**7 × 6 =**
12 × 2 =	**8 × 10 =**	**9 × 7 =**
9 × 2 =	**2 × 3 =**	**8 × 6 =**
9 × 6 =	**6 × 10 =**	**9 × 5 =**

GA1136

Multiplication Bingo

Object: To practice the multiplication facts 0 × 0 through 12 × 12. This game is for two or more players and works well with a large group.

Using the multiplication problems found on pages 60, 61 and 62, cut out the game cards. Shuffle the cards and place face down. Teacher or team player is called the "caller." Caller turns the top card up and announces the problem, but not the product. Each player looks to see if he has the appropriate product on his Bingo card. If the product is on his card, he scratches it off. The first player to get five products in a row scratched is declared the winner!

B	I	N	G	O
0, 1, 2, 3, 4, 5, 6, 7, 8, 9, 10, 11, 12, 13 **Write in any 5 of these, one in each square.**	16, 18, 20, 21, 22, 24, 25, 27, 28, 30, 32, 33, 35	36, 40, 42, 44, 45, 48, 49, 50	54, 55, 56, 60, 63, 64, 66, 70, 72, 77, 80, 81	84, 88, 90, 96, 99, 100, 108, 110, 120, 121, 132, 144

Choose any number listed under each letter to fill in the spaces in that column.

GA1136

Reproducible Facts Wheel
Multiplication

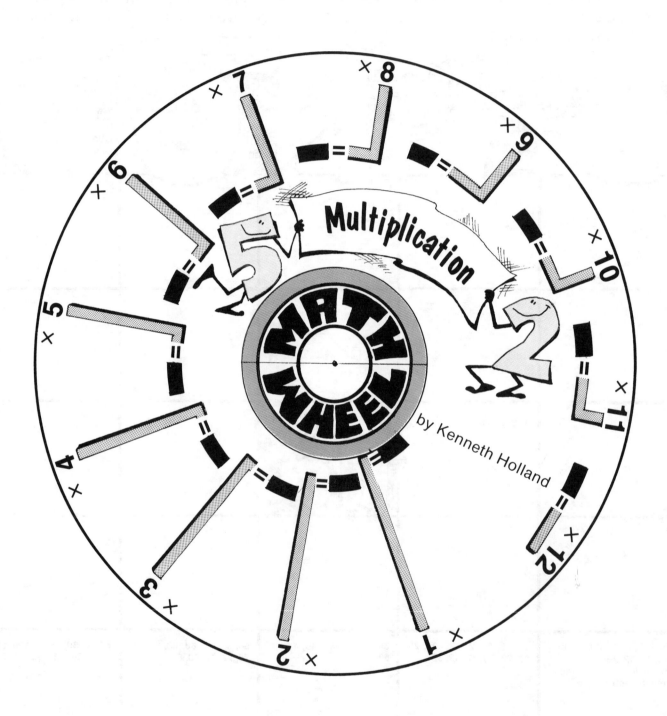

70

GA1136

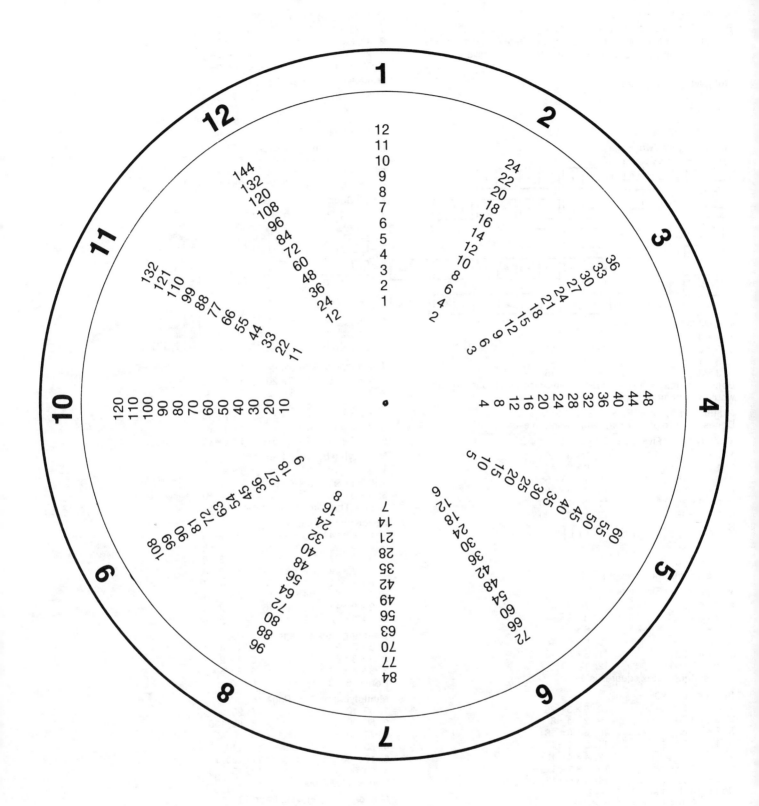

71

GA1136

Answer Key

Multiplying by Zero, Page 1
Bonus: Ten baskets with zero apples would have weight. Zero baskets of apples would not have weight. Therefore, ten baskets with zero apples would weigh more than zero baskets of ten apples.

Multiplying by One, Page 2
1. $2 \times 1 = 2$
2. $3 \times 1 = 3$
3. $1 \times 3 = 3$
4. $1 \times 27 = 27$

Multiplying by Two, Page 3

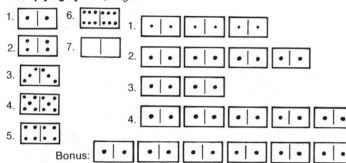

Multiplying by Three, Page 5
1. $2 \times 3 = 6$
2. $3 \times 2 = 6$
3. $3 \times 1 = 3$
4. $3 \times 3 = 9$
5. $1 \times 3 = 3$
6. $4 \times 3 = 12$
7. $4 \times 1 = 4$
8. $3 \times 5 = 15$
Bonus: Assuming the bags are the same size, one thousand bags would weigh more than 5 bags.

Multiplying by Four, Page 6
Bonus: Picture of four bags with twelve in each.

Multiplying by Five, Page 7
1. $2 \times 5 = 10$
2. $1 \times 5 = 5$
3. $3 \times 5 = 15$
4. $4 \times 5 = 20$
5. $6 \times 5 = 30$
6. $7 \times 5 = 35$
7. $8 \times 5 = 40$
8. $9 \times 5 = 45$
9. $11 \times 5 = 55$
10. $5 \times 5 = 25$
11. $10 \times 5 = 50$
12. $12 \times 5 = 60$
Bonus:

| | | | | | | | |
|||||||||
(five sets with four in each)

Multiplying by Six, Page 8
1. $1 \times 6 = 6$
2. $2 \times 6 = 12$
3. $4 \times 6 = 24$
4. $11 \times 6 = 66$
5. $5 \times 6 = 30$
6. $10 \times 6 = 60$
7. $7 \times 6 = 42$
8. $9 \times 6 = 54$
9. $6 \times 6 = 36$
10. $8 \times 6 = 48$
11. $3 \times 6 = 18$
12. $12 \times 6 = 72$
Bonus: Square. Six squares is 6×4.

Roll 'Em, Page 9
1. $2 \times 1 = 2$
2. $5 \times 6 = 30$
3. $2 \times 3 = 6$
4. $4 \times 5 = 20$
5. $5 \times 1 = 5$
6. $4 \times 6 = 24$
7. $3 \times 3 = 9$
8. $5 \times 5 = 25$
9. $2 \times 4 = 8$
10. $6 \times 6 = 36$
11. $2 \times 5 = 10$
12. $4 \times 4 = 16$
13. $1 \times 1 = 1$
14. $3 \times 5 = 15$
Bonus: 4: 1×4 and 2×2
6: 6×1 and 2×3
12: 6×2 and 3×4

Multiplication Check, Page 10
1. 10, 5 sets of 2 checks
2. 2, 1 set of 2 checks
3. 20, 4 sets of 5 checks
4. 15, 3 sets of 5 checks
5. 6, 3 sets of 2 checks
6. 12, 3 sets of 4 checks
7. 18, 3 sets of 6 checks
8. 24, 4 sets of 6 checks
9. 0, 0 sets of 5 checks
10. 12, 2 sets of 6 checks
11. 30, 5 sets of 6 checks
12. 0, 0 sets of 3 checks
13. 1, 1 set of 1 check
14. 24, 6 sets of 4 checks
15. 10, 2 sets of 5 checks
16. 8, 2 sets of 4 checks
Bonus: No, $2 \times 6 = 2$ sets of 6 checks,
$6 \times 2 = 6$ sets of 2 checks

Multiplying by Seven, Page 11
1. 7, 28, 56
2. 42, 63, 28
3. 21, 42, 14
4. 56, 7, 70
5. 35, 70, 63
6. 0, 35, 21
7. 77, 84, 0
8. 77, 84, 63

Multiplying by Eight, Page 12
1. 8, 40, 16
2. 48, 72, 0
3. 0, 24, 16
4. 40, 72, 56
5. 80, 8, 32
6. 96, 88, 56
7. 80, 32, 96
8. 24, 48, 88
Bonus: $8 = 8$, $16 = 1 + 6 = 7$, $24 = 2 + 4 = 6$, $32 = 3 + 2 = 5$, $40 = 4 + 0 = 4$, $48 = 4 + 8 = 12$, $56 = 5 + 6 = 11$, $64 = 6 + 4 = 10$, $72 = 7 + 2 = 9$, $80 = 8 + 0 = 8$, $88 = 8 + 8 = 16$, $96 = 9 + 6 = 15$

Magic Squares, Page 13

1.
0	6	0
4	■	2
0	0	0

2.
2	1	2
1	■	3
2	3	6

3.
2	2	4
2	■	2
4	2	8

4.
8	2	16
1	■	0
8	0	0

5.
4	2	8
3	■	0
12	0	0

6.
3	3	9
1	■	1
3	3	9

7.
5	1	5
2	■	4
10	2	20

8.
2	4	8
6	■	3
12	2	24

9.
3	4	12
2	■	1
6	2	12

The Shape of Things, Page 14
1. $7 \times 5 = 35$ $4 \times 5 = 20$ $3 \times 5 = 15$
2. $1 \times 1 = 1$ $4 \times 8 = 32$ $7 \times 7 = 49$
3. $2 \times 5 = 10$ $5 \times 6 = 30$ $1 \times 7 = 7$
4. $1 \times 2 = 2$ $1 \times 4 = 4$ $2 \times 7 = 14$
5. $2 \times 2 = 4$ $6 \times 2 = 12$ $6 \times 1 = 6$
6. $3 \times 7 = 21$ $5 \times 1 = 5$ $6 \times 7 = 42$

Multiplying by Nine, Page 15
1. $9 \times 2 = 18$ $9 \times 5 = 45$ $9 \times 8 = 72$
2. $9 \times 1 = 9$ $9 \times 7 = 63$ $9 \times 3 = 27$
3. $9 \times 6 = 54$ $9 \times 4 = 36$ $9 \times 9 = 81$
Bonus: The sum of each product is 9.

Decode and Multiply!, Page 16
1. $6 \times 4 = 24$ $3 \times 5 = 15$ $3 \times 3 = 9$
2. $7 \times 4 = 28$ $9 \times 4 = 36$ $5 \times 1 = 5$
3. $3 \times 9 = 27$ $4 \times 4 = 16$ $2 \times 3 = 6$
4. $7 \times 7 = 49$ $8 \times 3 = 24$ $4 \times 2 = 8$
5. $8 \times 7 = 56$ $9 \times 7 = 63$ $1 \times 1 = 1$
6. $1 \times 3 = 3$ $4 \times 1 = 4$ $6 \times 1 = 4$
7. $2 \times 9 = 18$ $8 \times 2 = 16$ $2 \times 2 = 4$
8. $6 \times 2 = 12$ $2 \times 1 = 2$ $7 \times 2 = 14$

Counting Cubes, Page 17
1. $3 \times 3 = 9$
2. $2 \times 3 = 6$
3. $3 \times 4 = 12$
4. $4 \times 4 = 16$
5. $4 \times 2 = 8$
6. $5 \times 5 = 25$
Bonus: $5 \times 3 = 15$ and $15 \times 4 = 60$

Multiplying by Ten, Page 18
1. 30, 10
2. 50, 20
3. 70, 80
4. 60, 70
5. 120, 50
6. 10, 60
7. 20, 90
8. 80, 100
9. 90, 40
10. 110, 30
11. 120, 110
12. 40, 120
Bonus: 11,000,000

Multiplying by Eleven, Page 19
1. 11, 22, 33
2. 44, 55, 66
3. 77, 88, 99
4. 110, 121, 132

Multiplying by Twelve, Page 20
1. 12, 36, 84
2. 84, 96, 24
3. 72, 60, 48
4. 48, 12, 36
5. 96, 24, 108
6. 60, 108, 120
7. 120, 144, 132
8. 72, 132, 0
Bonus: Same; each is 60.

GA1136

Multiplication Chart, Page 21

×	0	1	2	3	4	5	6	7	8	9	10	11	12
0	0	0	0	0	0	0	0	0	0	0	0	0	0
1	0	1	2	3	4	5	6	7	8	9	10	11	12
2	0	2	4	6	8	10	12	14	16	18	20	22	24
3	0	3	6	9	12	15	18	21	24	27	30	33	36
4	0	4	8	12	16	20	24	28	32	36	40	44	48
5	0	5	10	15	20	25	30	35	40	45	50	55	60
6	0	6	12	18	24	30	36	42	48	54	60	66	72
7	0	7	14	21	28	35	42	49	56	63	70	77	84
8	0	8	16	24	32	40	48	56	64	72	80	88	96
9	0	9	18	27	36	45	54	63	72	81	90	99	108
10	0	10	20	30	40	50	60	70	80	90	100	110	120
11	0	11	22	33	44	55	66	77	88	99	110	121	132
12	0	12	24	36	48	60	72	84	96	108	120	132	144

Arrow Multiplication, Page 23

1. $1 \times 2 = 1$
2. $1 \times 4 = 4$
3. $5 \times 4 = 20$
4. $6 \times 7 = 42$
5. $8 \times 1 = 8$
6. $9 \times 8 = 72$
7. $3 \times 4 = 12$
8. $4 \times 3 = 12$
9. $7 \times 6 = 42$
10. $4 \times 5 = 20$
11. $3 \times 2 = 6$
12. $6 \times 5 = 30$
13. $7 \times 8 = 56$
14. $9 \times 2 = 18$
15. $2 \times 3 = 6$

Bonus: ↖4, 6↓ 8↓, ↑1, ↓2, 4↗

Product Dot-to-Dot, Page 25

1. 12
2. 18
3. 24
4. 25
5. 30
6. 20
7. 21
8. 36
9. 81
10. 100
11. 35
12. 44
13. 144
14. 88
15. 120
16. 45
17. 54

Computer Multiplication, Page 26

1. 12
2. 22
3. 20
4. 34
5. 28
6. 46
7. 36
8. 58
9. 44
10. 70
11. 52
12. 82

Bonus:
1. 16
2. 16
3. 28
4. 24
5. 40
6. 32
7. 52
8. 40
9. 64
10. 48
11. 76
12. 56

Coded Multiplication, Page 27

1. $4 \times 11 = 44$ $9 \times 7 = 63$ $12 \times 5 = 60$
2. $5 \times 5 = 25$ $5 \times 7 = 35$ $12 \times 7 = 84$
3. $6 \times 7 = 42$ $11 \times 7 = 77$ $12 \times 6 = 72$
4. $8 \times 7 = 56$ $9 \times 5 = 45$ $7 \times 10 = 70$
5. $8 \times 8 = 64$ $9 \times 6 = 54$ $12 \times 4 = 48$
6. $8 \times 9 = 72$ $8 \times 10 = 80$ $7 \times 7 = 49$
7. $10 \times 5 = 50$ $10 \times 12 = 120$ $11 \times 12 = 132$
8. $10 \times 10 = 100$ $10 \times 6 = 60$ $5 \times 11 = 55$
9. $9 \times 12 = 108$ $9 \times 9 = 81$ $8 \times 12 = 96$
10. $3 \times 5 = 15$ $3 \times 3 = 9$ $8 \times 11 = 88$
11. $8 \times 6 = 48$ $12 \times 12 = 144$ $3 \times 2 = 6$
12. $11 \times 6 = 66$ $3 \times 1 = 3$ $11 \times 2 = 22$

Multiplication Towers, Page 28

1.
```
          0
        0 0
      0 0 0
   12 0 0 0
    2 6 0 0 2
   1 2 3 0 1 2
```

2.
```
        0 32 0
      0 4 8 0
    0 2 2 4 0
  0 2 1 2 2 0
```

3.
```
        1,048,576
      0 1024 1024 0
    0 16 64 16 0
  0 2 8 8 2 0
 0 1 2 4 2 1 0
0 1 1 2 2 1 1 0
```

Search and Circle, Page 29

What's Your Sign?, Page 30

1. $8 \times 2 = 16$ $6 \times 11 = 66$
2. $11 \times 12 = 132$ $120 = 10 \times 12$
3. $2 \times 7 = 14$ $1 \times 10 = 10$
4. $8 \times 6 = 48$ $7 \times 11 = 77$
5. $7 \times 10 = 70$ $12 \times 1 = 12$
6. $6 \times 6 = 36$ $22 = 11 \times 2$
7. $12 = 6 \times 2$ $1 \times 11 = 11$
8. $8 \times 8 = 64$ $5 \times 7 = 35$
9. $20 = 10 \times 2$ $5 \times 8 = 40$
10. $12 \times 2 = 24$ $12 \times 8 = 96$
11. $3 \times 6 = 18$ $80 = 10 \times 8$
12. $4 \times 3 = 12$ $8 \times 9 = 72$
13. $10 \times 5 = 50$ $5 \times 9 = 45$
14. $6 \times 7 = 42$ $5 \times 11 = 55$
15. $6 \times 9 = 54$ $5 \times 12 = 60$
16. $12 \times 12 = 144$ $10 \times 6 = 60$
Bonus: $120 \times 0 = 0$

More Multiplication Magic Squares, Page 31

1.

3	3	9
2	■	4
6	6	36

2.

2	5	10
4	■	4
8	5	40

3.

5	3	15
4	■	4
20	3	60

4.

4	2	8
1	■	3
4	6	24

5.

5	2	10
20	■	10
100	1	100

6.

6	2	12
10	■	10
60	2	120

Bonus:

5	2	10
5	■	10
25	4	100

Perfectly Preposterous Problems, Page 32

1. $7 \times 6 = 42$
2. $4 \times 5 = 20$
3. $10 \times 3 = 30$
4. $9 \times 0 = 0$
5. $2 \times 12 = 24$
Bonus: $8 \times 7 = 56$ $56 \times 3 = 168$

"X" Marks the Problem, Page 33

1. $10 \times 12 = 120$ $12 \times 6 = 72$ $9 \times 9 = 81$
2. $8 \times 9 = 72$ $10 \times 10 = 100$ $10 \times 6 = 60$
3. $8 \times 8 = 64$ $12 \times 5 = 60$ $8 \times 12 = 96$
4. $7 \times 7 = 49$ $7 \times 12 = 84$ $5 \times 12 = 60$
5. $8 \times 6 = 48$ $7 \times 9 = 63$ $10 \times 9 = 90$
6. $9 \times 11 = 99$ $7 \times 5 = 35$ $8 \times 11 = 88$
7. $9 \times 6 = 54$ $12 \times 12 = 144$ $12 \times 4 = 48$
8. $7 \times 6 = 42$ $11 \times 10 = 110$ $11 \times 12 = 132$
Bonus: Answers will vary. $1 \times 12 = 12$, $3 \times 4 = 12$
$4 \times 3 = 12$, $6 \times 2 = 12$, $12 \times 1 = 12$, $2 \times 6 = 12$

Multiplication Connection, Page 34

$1 \times 11 = 11$ $1 \times 7 = 7$
$2 \times 3 = 6$ $2 \times 12 = 24$
$3 \times 6 = 18$ $3 \times 9 = 27$
$4 \times 7 = 28$ $4 \times 11 = 44$
$5 \times 9 = 45$ $5 \times 4 = 20$
$6 \times 12 = 72$ $6 \times 6 = 36$
$7 \times 10 = 70$ $7 \times 5 = 35$
$8 \times 8 = 64$ $8 \times 10 = 80$
$9 \times 5 = 45$ $9 \times 8 = 72$
$10 \times 4 = 40$ $10 \times 2 = 20$
$11 \times 2 = 22$ $11 \times 1 = 11$
$12 \times 1 = 12$ $12 \times 3 = 36$

Secret Message, Page 35

You are very special.

GA1136

Multiplication Crossword, Page 37

Across:
1. 3 × 7 = 21
2. 10 × 2 = 20
3. 5 × 6 = 30
4. 12 × 2 = 24
5. 9 × 10 = 90
6. 5 × 8 = 40
7. 10 × 3 = 30
8. 10 × 8 = 80
9. 7 × 6 = 42
10. 3 × 12 = 36
11. 1 × 1 = 1
12. 10 × 9 = 90
13. 8 × 11 = 88
14. 12 × 8 = 96
15. 3 × 12 = 36
16. 3 × 8 = 24
17. 7 × 7 = 49
18. 1 × 12 = 12

Down:
1. 4 × 5 = 20
2. 10 × 2 = 20
3. 10 × 3 = 30
4. 5 × 4 = 20
5. 10 × 9 = 90
6. 6 × 7 = 42
7. 9 × 4 = 36
8. 8 × 10 = 80
9. 8 × 6 = 48
10. 12 × 3 = 36
11. 2 × 8 = 16
12. 8 × 12 = 96
13. 7 × 12 = 84
14. 9 × 11 = 99
15. 8 × 4 = 32

Multiplication Paths, Page 38

1. 12 × 3 = 36
2. 8 × 12 = 96
3. 10 × 11 = 110
4. 9 × 9 = 81
5. 12 × 7 = 84
6. 11 × 9 = 99
7. 7 × 7 = 49
8. 8 × 8 = 64
9. 9 × 12 = 108
10. 12 × 12 = 144

Bull's-Eye Multiplication, Page 39

1. 12 × 10 = 120
   12 × 8 = 96
   12 × 9 = 108
   12 × 5 = 60
   12 × 6 = 72
   12 × 7 = 84
   12 × 4 = 48
   12 × 3 = 36
2. 9 × 9 = 81
   9 × 3 = 27
   9 × 7 = 63
   9 × 6 = 54
   9 × 8 = 72
   9 × 4 = 36
   9 × 5 = 45
   9 × 10 = 90

3. 10 × 6 = 60
   10 × 5 = 50
   10 × 7 = 70
   10 × 3 = 30
   10 × 8 = 80
   10 × 9 = 90
   10 × 4 = 40
   10 × 10 = 100
4. 11 × 9 = 99
   11 × 6 = 66
   11 × 7 = 77
   11 × 4 = 44
   11 × 8 = 88
   11 × 5 = 55
   11 × 10 = 110
   11 × 3 = 33

Multiplication Boggler, Page 41

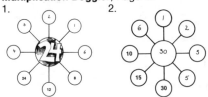

Full Circle Multiplication, Page 42

1. 12 × 3 = 36
   11 × 8 = 88
   9 × 4 = 36
   10 × 9 = 90
2. 9 × 9 = 81
   7 × 8 = 56
   8 × 9 = 72
   12 × 7 = 84
3. 12 × 12 = 144
   12 × 11 = 132
   12 × 9 = 108
   10 × 12 = 120
4. 5 × 12 = 60
   7 × 11 = 77
   11 × 10 = 110

Bonus: 120 × 0 = 0

Circle Three, Page 43

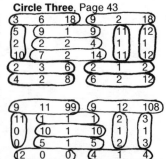

Product Designs, Page 44

1. 6: 1 × 6, 2 × 3
2. 10: 1 × 10, 2 × 5
3. 9: 1 × 9, 3 × 3
4. 12: 1 × 12, 2 × 6, 3 × 4
5. 24: 1 × 24, 2 × 12, 3 × 8, 4 × 6
Bonus: 36: 1 × 36, 2 × 18, 3 × 12, 4 × 9, 6 × 6

Product Pairs, Page 45

1. 20: 1 × 20, 2 × 10, 4 × 5
2. 36: 1 × 36, 2 × 18, 3 × 12, 4 × 9, 6 × 6
3. 12: 1 × 12, 2 × 6, 3 × 4
4. 100: 1 × 100, 2 × 50, 4 × 25, 10 × 10, 5 × 20
5. 120: 2 × 60, 4 × 30, 5 × 24, 10 × 12
6. 40: 2 × 20, 4 × 10, 5 × 8
Bonus: 1: 1 × 1 = 1

Product Matchup, Page 46

1. 3 × 6 = 2 × 9
2. 9 × 0 = 12 × 0
3. 1 × 4 = 2 × 2
4. 1 × 6 = 2 × 3
5. 10 × 1 = 5 × 2
6. 3 × 4 = 2 × 6
7. 5 × 4 = 10 × 2
8. 6 × 4 = 8 × 3
9. 5 × 8 = 4 × 10
10. 8 × 1 = 2 × 4
11. 8 × 2 = 4 × 4
12. 6 × 8 = 12 × 4
13. 6 × 6 = 9 × 4
14. 5 × 6 = 10 × 3
Bonus: EVERY DAY

Hidden Number Pairs, Page 47

Consecutive Products, Page 49

1. 2, 3
2. 4, 5
3. 0, 1
4. 3, 4
5. 9, 10
6. 5, 6
7. 10, 11
8. 8, 9
9. 7, 8
10. 12, 13
Bonus: 0, 1 and 1, 2

Multiplication Patterns, Page 50

1. 14, 14, 16—add 2, multiply by 1, repeat, etc.
2. 4, 8, 8—multiply by 1, multiply by 2, repeat, etc.
3. 22, 44, 46—add 2, multiply by 2, repeat
4. 74, 148, 158—add 10, multiply by 2, repeat, etc.
5. 10, 20, 10—subtract 10, multiply by 2, repeat, etc.
6. 120, 720, 5760—multiply by 1, 2, 3, 4, 5, 6, 7, 8, etc.
Bonus: T, E, T, T, F, etc. The first letter of each number word is used.

Get This Straight, Page 51

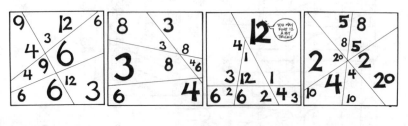

Three Straight Lines, Page 52

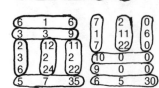

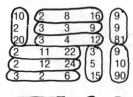

GA1136

Making Arrangements, Page 53

1.	423 × 1 423	5.	341 × 2 682	9.	321 × 4 1284
2.	431 × 2 862	6.	314 × 2 628	10.	312 × 4 1248
3.	143 × 2 286	7.	412 × 3 1236	11.	342 × 1 342
4.	432 × 1 432	8.	421 × 3 1263	12.	124 × 3 372

Mystery Digits, Page 54

1. 5, 2. 6, 3. 2, 4. 1, 5. 9

Bonus: 0, 2

Multiplication Magic, Page 55

The product in step 3 will be the number chosen in step 1, repeated many times.

Bonus: The product in step 3 will be the number chosen in step 1, repeated many times, except there will be a 0 in the ten's place.

Magical Eleven, Page 56

1. 121
2. 12321
3. 1234321
4. 123454321
5. 12345654321
6. 1234567654321
7. 123456787654321

Bonus: 12345678987654321

Dropping Zeros, Page 57

The answer in step 8 will be the same as the number chosen in step 1.

Pick Three, Any Three, Page 58

The square of the middle number will be one less or one more than the product of the other two numbers.

Bonus: Each number is the sum of the two previous listed numbers. To extend the pattern, add 89, 144, 233, 377, etc.

Calculator Magic, Page 59

1. 88
2. 888
3. 8888
4. 88888
5. 888888
6. 8888888
7. 88888888
8. 888888888

Bonus:

1. 11
2. 111
3. 1111
4. 11111
5. 111111
6. 1111111
7. 11111111
8. 111111111

Pre/Post Test, Page 60

2 × 1 = 2	10 × 8 = 80	7 × 0 = 0
2 × 3 = 6	8 × 6 = 48	7 × 9 = 63
3 × 5 = 15	7 × 1 = 7	11 × 5 = 55
2 × 0 = 0	4 × 6 = 24	8 × 7 = 56
7 × 4 = 28	3 × 1 = 3	4 × 2 = 8
11 × 0 = 0	12 × 4 = 48	10 × 7 = 70
10 × 2 = 20	10 × 5 = 50	11 × 3 = 33
10 × 1 = 10	3 × 0 = 0	11 × 8 = 88
3 × 10 = 30	11 × 7 = 77	12 × 10 = 120
9 × 9 = 81	11 × 12 = 132	12 × 12 = 144

Pre/Post Test, Page 61

3 × 3 = 9	8 × 0 = 0	8 × 1 = 8
12 × 9 = 108	10 × 4 = 40	7 × 2 = 14
10 × 10 = 100	4 × 9 = 36	8 × 3 = 24
2 × 6 = 12	7 × 6 = 42	4 × 4 = 16
11 × 9 = 99	12 × 2 = 24	5 × 6 = 30
12 × 8 = 96	3 × 7 = 21	8 × 4 = 32
12 × 11 = 132	7 × 5 = 35	4 × 0 = 0
9 × 0 = 0	5 × 1 = 5	11 × 6 = 66
10 × 6 = 60	9 × 6 = 54	12 × 1 = 12
8 × 8 = 64	12 × 7 = 84	12 × 5 = 60

Pre/Post Test, Page 62

1 × 1 = 1	2 × 2 = 4	9 × 2 = 18
11 × 11 = 121	6 × 6 = 36	10 × 9 = 90
4 × 3 = 12	9 × 8 = 72	6 × 1 = 6
11 × 4 = 44	2 × 8 = 16	5 × 4 = 20
1 × 0 = 0	6 × 3 = 18	10 × 11 = 110
5 × 5 = 25	5 × 2 = 10	6 × 0 = 0
10 × 0 = 0	12 × 6 = 72	7 × 7 = 49
4 × 1 = 4	9 × 1 = 9	9 × 3 = 27
11 × 2 = 22	5 × 9 = 45	11 × 1 = 11
5 × 0 = 0	12 × 3 = 36	12 × 0 = 0

Page 67

12 × 12 = 144	11 × 3 = 33	9 × 9 = 81
8 × 5 = 40	2 × 11 = 22	7 × 2 = 14
12 × 6 = 72	9 × 3 = 27	7 × 5 = 35
6 × 6 = 36	11 × 12 = 132	11 × 8 = 88
12 × 5 = 60	4 × 5 = 20	5 × 5 = 25
6 × 4 = 24	7 × 7 = 49	11 × 6 = 66
7 × 3 = 21	9 × 10 = 90	6 × 3 = 18
10 × 12 = 120	4 × 4 = 16	11 × 11 = 121
12 × 8 = 96	6 × 5 = 30	8 × 2 = 16
5 × 3 = 15	12 × 0 = 0	6 × 2 = 12

Page 68

10 × 11 = 110	12 × 4 = 48	9 × 11 = 99
8 × 9 = 72	12 × 3 = 36	7 × 4 = 28
9 × 4 = 36	10 × 3 = 30	11 × 4 = 44
8 × 4 = 32	2 × 2 = 4	5 × 2 = 10
8 × 3 = 24	7 × 8 = 56	11 × 2 = 22
3 × 4 = 12	10 × 10 = 100	8 × 8 = 64
9 × 12 = 108	12 × 7 = 84	7 × 6 = 42
12 × 2 = 24	8 × 10 = 80	9 × 7 = 63
9 × 2 = 18	2 × 3 = 6	8 × 6 = 48
9 × 6 = 54	6 × 10 = 60	9 × 5 = 45

GA1136

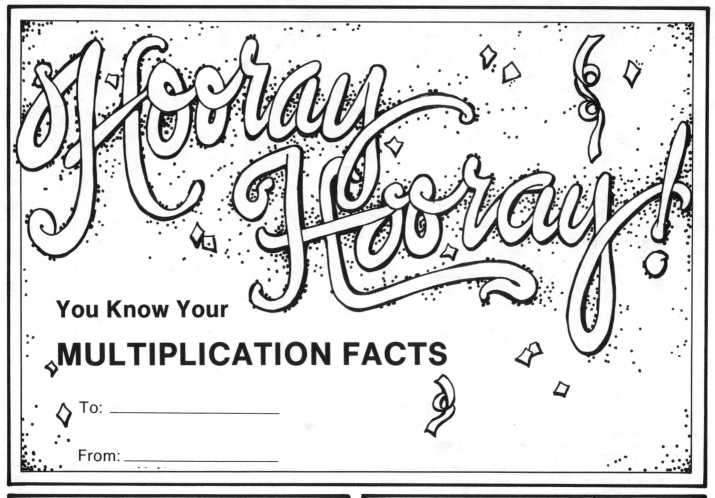

You Know Your

MULTIPLICATION FACTS

To: _____

From: _____

X-traordinary at X-ing!*

*multiplication, not kissing!

To: _____

From: _____

XXXXXXX

STAR OF MATH FACTS

To: _____

From: _____

76

GA1135

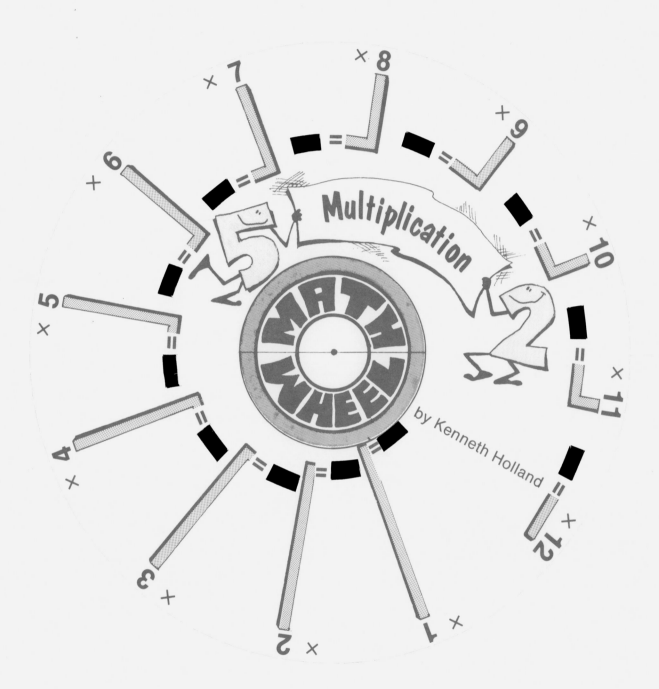

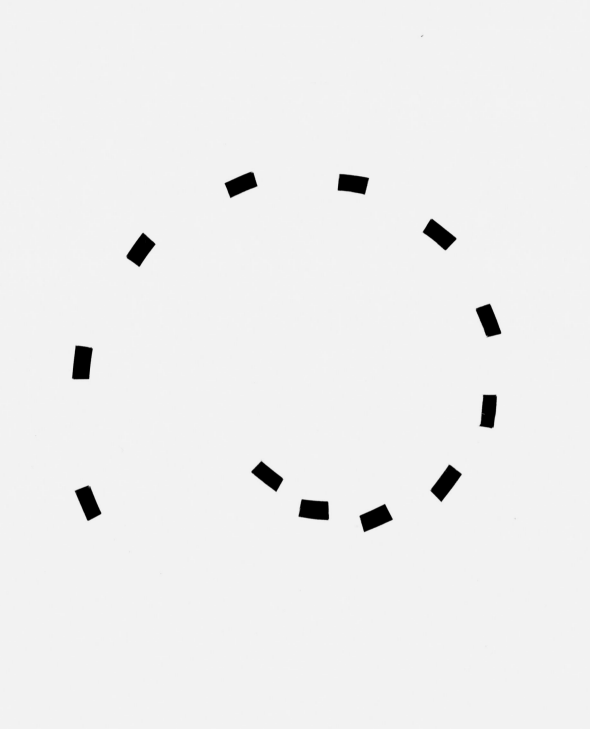